JOURNEY INTO THE NEW COSMOLOGY

JOURNEY INTO THE NEW COSMOLOGY
A Scientific and Mystical Exploration

David P. Judd

Authors Choice Press
San Jose New York Lincoln Shanghai

Journey Into the New Cosmology
A Scientific and Mystical Exploration

All Rights Reserved © 2000 by David P. Judd

No part of this book may be reproduced or transmitted in any form or by any means, graphic, electronic, or mechanical, including photocopying, recording, taping, or by any information storage or retrieval system, without the permission in writing from the publisher.

Authors Choice Press
an imprint of iUniverse.com, Inc.

For information address:
iUniverse.com, Inc.
5220 S 16th, Ste. 200
Lincoln, NE 68512
www.iuniverse.com

All Biblical references are to the Revised Standard Version (New York: Oxford University Press, 1962)

ISBN: 0-595-15218-X

Printed in the United States of America

Contents

Introduction		...ix
Chapter One	What Is a Cosmology?	...1
Chapter Two	Quantum Physics	...10
Chapter Three	Epistemology: How Do We Know?	...19
Chapter Four	The New Cosmology	...30
Chapter Five	A Nonphysical World	...41
Chapter Six	A New Theology	...52
Chapter Seven	Love and Meaning	...60
Chapter Eight	Medicine and Psychology	...66
Chapter Nine	Personal Application and Conclusion	...75
About the Author		...81
Notes		...83

Acknowledgements

I have read many books, heard many songs, and encountered many people throughout my life that have inspired me and contributed to my thought, understanding and experience. All have impacted my development at various stages of growth. I express my appreciation to all these people whether I have personally met them or just benefited from their work. I would specifically like to thank my editor Jessica Palmer and my wife Allison for their input and encouragement in the writing of this book.

Introduction

This book is about the thought and research that is radically changing our understanding of the world we live in. It is truly nothing less than a revolution. We are moving away from the idea that natural laws or forces outside of ourselves control our lives and are coming into a new appreciation of the creative power of individual thought. The evidence is accumulating in nearly every academic discipline and field of endeavor, and there are many books on the market today that detail this evidence for each specific field. The purpose of my writing is not to reiterate all this evidence, but to examine why our thoughts affect the world of our physical experience in the way they do. To understand the "why" of it, we have to delve deep down into the basic assumptions we have about reality. We have to look at our cosmology.

To continue to progress, we need a new model for understanding ourselves and the world of our experience because the classical model simply cannot explain much of the evidence accumulating today. The new cosmology helps us to understand and take control of our experience in a way that was not possible before. In developing a picture of the reality being conceived today, I draw upon a wide variety of evidence from quantum physics to mysticism. My attempt is to provide a wide sampling of evidence from many different fields, not to exhaust the evidence in any one field. My attempt also is to bring this information and thought together in a way that is understandable to the general public.

The implications of this new cosmology for medicine, psychology, theology and virtually every field of inquiry are quite astounding and are explored in these pages. We have begun a new journey. It promises to be an exciting one.

David Judd

Chapter One
What Is a Cosmology?

A "cosmology" is the basic set of assumptions we have about the world in which we live. It is the underlying framework of understanding from which we interpret our existence; it informs us of the essential nature and basic laws of the reality we experience. Western culture has been dominated the last couple of centuries by the assumptions of classical or "Newtonian" physics. This cosmology asserts that we live in a mechanistic reality controlled by natural laws and says this reality is composed of solid particles called "matter." Classical cosmology asserts other basic principles, all of which have been well accepted for so long that most people consider them indisputable facts. It is thus absolutely astonishing that quantum physicists are today stating that these "facts" are not true, and that reality is actually quite different from what we previously thought.

In this book we will explore the discoveries of quantum physics as well as related insights from various academic and mystical sources. We will develop a picture of the new cosmology that is unfolding in our time, a cosmology that has profoundly significant, exciting and positive implications. To more fully appreciate this picture, however, there are some preliminary issues to be addressed first.

The Cosmological Box

Human beings naturally want to explore their world and understand it. This is why we develop cosmologies, that basic set of ideas about what the world is and how it works. A problem arises, however, when we begin to see our cosmological perspective as *fact*, and not merely assumption. When this happens the growth in our understanding becomes stunted; we cannot learn because we think we already know. We then find ourselves in a conceptual box that is difficult to see outside of because all the evidential data we encounter is interpreted in a way that will make it fit into the box. If necessary, we twist the data around, mold it, and bend it out of shape until it fits our box; evidence that does not fit is either ignored or discounted. Our conceptual boxes thus becomes self-validating and self-reinforcing. We do not see that many of our explanations are incomplete and contrived or that there are better explanations for some things. This is not a conscious or deliberate process. We do it because we do not recognize that the parameters of the box can or ought to be questioned. We do it because there is a sense of false security when we live in a box.

There is a great deal of evidence that suggests some kind of nonphysical level of existence. There are many reported instances of psychic visions, near-death experiences, ESP, and so forth. The classical cosmologist ignores, discounts or explains away this evidence because it does not fit into their box. They say it is only the imagination, or the person is mentally unbalanced, or it was caused by some chemical or hormone, or they just ignore it altogether. The discoveries of quantum physics make it clear these kinds of phenomena ought to be seriously explored, but the main point here is to recognize how our conceptual boxes exert a determining influence on our interpretations of evidential data.

The self-validating nature of beliefs can be observed on an ordinary, everyday level. If people believe themselves to be unworthy and unlovable, they will interpret the actions and responses of others in a way that will support their belief. If someone is angry with them, they tend to interpret this to mean that there is something wrong with them. On the other hand, if people believe themselves worthy and lovable, they interpret the actions and responses of others differently. If someone is angry they say he or she is having a bad day or some other such thing. The observed evidential data is the same in both cases. The *interpretation* of it is based on preexisting assumptions or beliefs. Our evaluations of others are likewise influenced by what we believe about them personally, or by what we believe about people in general. If, for example, we do not believe people can be trusted, we look for an ulterior motive in any act of seeming kindness. The self-validating nature of beliefs is expressed well in *A Course in Miracles* when it says, "Perception is a mirror."[1]

Another reason it is difficult to see outside a conceptual box is language. Language is developed to express the concepts we now have, whether that be of an ordinary physical object like "tree," or something quite abstract. We do not have language to express those things of which we have not yet conceived. People who see outside of the box have a hard time communicating what they see because they have to use the old language. They try to get creative about it by using paradox, analogy, and so on, but it is still difficult. The challenge of language is one that quantum physicists and others have had to face as they have tried to talk about a cosmological picture of reality that is different from the classical one.

It can be difficult to see beyond our self-created conceptual boxes. To do so we begin by acknowledging that the things we consider to be facts are no more than assumptions, and may not be true. We recognize what these assumptions are and then question them. It may not be easy to question these assumptions because, in fact, we live in a cultic society. The term "cult" is not simply a reference to some off-the-wall religious group; it actually refers to any group, organization or institution in which the members are required to believe or accept certain ideas, beliefs or facts without question. The more facts that cannot be questioned within a group, the more cultic the group.

Even a cursory examination reveals that cultism is woven into the fabric of some of our most revered societal institutions. In the field of medicine, for example, we were formerly thought to be just bodies. Every physical symptom was thought to have a strictly physical cause: chemicals, bacteria, viruses and the like. Practitioners who thought the mind could play a part in illness or who considered alternative methods of treatment were ridiculed; their voices were squelched and, if they persisted in their "nonsense," were cast out of the medical society or school. The field of medicine is today opening up to new views but we still must ask, what are the new or remaining cultic limitations? In what ways do those in power still force their views on others and squash open inquiry?

The field of psychology also has had a limited view of who we are as human beings that it has forced upon its students and practitioners. Psychology acknowledges our mental and emotional experience but still considers them to be *physically* based. Psychological problems are genetically or environmentally determined; they are effects of the physical brain, experiences in the physical world, or chemicals or hormones in the physical body. The explanations have all been physical in one sense or another. The possibility that humans might be more than physical, or might have a spiritual or nonphysical aspect, could not be raised. Today this too is changing as sub-disciplines such as Transpersonal Psychology are gaining more recognition, but progress has been slow.

Academia in general has been a bastion of cultism. Mainstream academia has only accepted interpretations of data that are consistent with the currently dominant cosmological assumptions. We can readily acknowledge that the explanations offered to us by the various academic disciplines make sense. It must be recognized, however, that "making sense" is simply the process of being consistent within the confines of one's assumptions. If the only thing considered to be real is the physical world, then every observed effect is found to have a physical cause or explanation. All the incoming data is made to fit that model or simply ignored. The unquestioned assumptions thus *determine* the eventual conclusions which, in turn, validate the assumptions. Perhaps this becomes clearer when we look at an extreme example. White supremacists begin with the assumption that white Protestants are superior to blacks, Jews and Catholics, and interpret data accordingly. Data that supports their assumptions is, of course, readily accepted. Data that does not is either rejected or twisted around and reinterpreted until it does fit. We acknowledge that this is not a matter of intelligence; many white supremacists would do well on standardized intelligence tests. Instead, it is the matter of not questioning one's basic assumptions.

Despite many dissenting voices and much evidence to the contrary, our culture has consistently refused to question the basic assumptions of classical physics. Fortunately for all of us our most revered institution, science itself, is now doing the questioning. This gives us an unprecedented opportunity to see outside our conceptual box.

The Influence of Cosmology on Theology

Every field of inquiry is influenced by the basic set of assumptions we call a cosmology, even theology. Some examples might help make this clearer. In biblical times people had an understanding of the physical world that would sound strange to us today. The earth was believed to be flat and surrounded by mountains upon which rested a solid physical arch called the "firmament" (Genesis, 1:6, ff.). There were waters above this solid arch that were let down through windows in the form of rain. The stars were set in the firmament like so many precious stones. The firmament was also called heaven and was the location of God's throne. Hell was an actual physical place under the earth. God was near in this cosmology; I mean, how far could He go? This was an extremely small universe by today's standards.

As humanity's understanding of the physical universe expanded, we realized that space was actually immense and that there was not a solid arch over the earth. We recognized that the earth was not the center of the universe and that the sun did not revolve around it. These new understandings precipitated a crisis for religious people. Where was God in this vast universe and how significant could humans be? Religious people fought against these new understandings of the physical universe for their theology was tied to the old understandings. The Church condemned the discoveries of Galileo. But theology adapted and changed from the physically oriented conceptions of that time to the more spiritual or nonphysical conceptions of our own time. Today, mainline Christians do not believe God physically resides in the "firmament," and do not believe that Hell, if they believe in it at all, is an actual physical place under the earth. Changes in cosmological perspectives brought changes to theological perspectives.

Another example is found in the development of the theory of evolution, which created another theological crisis. The *Interpreter's Bible* states:

> "That collision between the new teachings and the old tradition seemed at first as shattering as an earthquake. Multitudes of men and women reacted in panic or in defiance, supposing that if their confidence in the literal exactitude of the first verses of Genesis should go, then their whole religious faith would be gone with it. Yet the new teachings had come not to blight religion but to stimulate it to new growth. If they seemed at first to break up old patterns of belief, the result was to lift

men's eyes to mightier perspectives of the majestic works of God."²

Theology adapts as new understandings of the physical world emerge. When the theory of evolution became popular, many Christian theologies moved away from the idea that God created the world in six days and developed other understandings of God's creative act. It is natural for theology to develop in this way. In fact, it is impossible for it to do otherwise. In thinking about the Divine, or anything else for that matter, we naturally develop ideas based on our current concepts. Our thoughts, ideas and perceptions are limited to the confines of our conceptual boxes.

The word "paradigm" is often used these days in talking about the conceptual box. A related word that theologians sometimes use is "anthropomorphism." A humorous way of understanding what is meant by anthropomorphism is the suggestion that "man created God in his own image." When God is thought of as angry, vengeful or condemning as written in the Old Testament, this is considered by many to be anthropomorphic. The God they conceived of told them to slaughter other peoples, sometimes unmercifully. The God they conceived of thought justice was a matter of an "eye for an eye." The concepts of the new cosmology provide some exciting new possibilities for theology.

Authority

Another important part of the process of stepping outside our conceptual boxes has to do with an understanding of what constitutes authority on any matter under consideration. How do we know what is true? Who is the authority? We can use religion as an example, but the ideas apply equally to any other field of inquiry. In matters of religion some people say the Bible is the authority; others might say the Pope or some other religious leader is the authority. Yet others might cite some particular scholar, theologian or tradition. How does one choose the correct authority when there are almost as many interpretations of the Bible as there are people, and for every Ph.D. or authority on one side of an issue, there is one on the other side?

The inescapable fact is that each person is his or her own authority. Each person has to decide which authority or interpretation is the correct one. People are, of course, limited in their ability to decide by the extent of their exposure to different ideas and different life experiences. They are limited by such things as the fear of new ideas, the influence of others and intellectual ability. Nevertheless, we each decide what seems right and makes sense to us, and we acknowledge the "authorities" that concur with that perception.

There are so many arguments about what is right and what the truth is and everyone quotes his or her own sources as if this would prove anything, but there is no authority that everyone can agree is authoritative. The sources we consider authoritative are simply *the ones we agree with*! Again, it comes back to us. The bottom line is *what* is said, not who said it.

There is one more issue to be raised in regard to the subject of authority. Those in power, whether we are talking about religion or secular academia, tend to be conservators of the status quo who want to retain the beliefs and conclusions now in place. Some call it a natural human tendency for those in power to want to stay in power. Whether this tendency is "natural" or not, it seems clear that people in power often feel threatened by new ideas and fight against them. The individual must take this into consideration when weighing the opinions of authorities.

As we evaluate the precepts of a new cosmology, we could begin by evaluating the relative merit of the authorities involved. We will, however, find respectable, intelligent people on all sides of any issue raised. In the final analysis, therefore, it will come down to what we personally think and feel. In my view, the explanations for experience offered by the new cosmology are more reasonable and comprehensive than the classical ones.

Chapter Two
Quantum Physics

Quantum physics does not merely rearrange or adjust the principles of classical physics, it undermines them so completely that the term "physical" is given a whole new meaning. It is as if the whole idea of a physical object as understood in the classical sense does not, and never did, exist as such! In this regard, Fritjof Capra quotes the pre-eminent quantum physicist Werner Heisenberg and Albert Einstein as follows:

Heisenberg: "The violent reaction on the recent development of modern physics can only be understood when one realises (sic) that here the foundations of physics have started moving; and that this motion has caused the feeling that the ground would be cut from science."
Einstein: "All my attempts to adapt the theoretical foundation of physics to this (new type of) knowledge failed completely. It was as if the ground had been pulled out from under one, with no firm foundation to be seen anywhere, upon which one could have built."[3]

Capra himself states, "The first three decades of our century changed the whole situation in physics radically. Two separate developments—that of relativity theory and of atomic physics—shattered all the principal concepts of the Newtonian world view: the notion of absolute space and time, the elementary solid particles, the strictly causal nature of physical phenomena, and the ideal of an objective description of nature."[4] Quantum physics cannot even be adequately talked about in the terminology of classical physics, but since that is the only language we have, we just do the best we can. Another problem with understanding quantum physics is our tendency to try and visualize something to understand it, but visualization may not be possible with quantum physics. Gary Zukav states: "Atoms were never 'real' things anyway. Atoms are hypothetical entities constructed to make experimental observations intelligible. No one, not one person, has ever seen an atom. Yet we are so used to the *idea* that an atom is a thing that we forget that it is an idea. Now we are told that not only is an atom an idea, it is an idea that we cannot even picture."[5]

The experimental results of quantum physics may seem like nonsense to persons familiar only with classical thought. Perhaps we only need be reminded of the "nonsense" promulgated by scientists such as Copernicus who insisted the earth revolved around the sun and not vice versa. I like Zukav's definition: "'Nonsense' is that which does not fit into the prearranged patterns which we have super-imposed on reality."[6]

Perhaps what is required to understand quantum physics is a beginner's mind. Zukav quotes Baker Roshi, the American Zen master, as saying, "'The mind of the beginner is empty, free of the habits of the expert, ready to accept, to doubt, and open to all the possibilities.'"[7] Perhaps another way of dealing with the discoveries of quantum physics is through the concept of paradox. Webster's defines "paradox" as: "1. A statement contrary to common belief. 2. A statement that seems contradictory, unbelievable, or absurd but that may actually be true in fact."[8] Zukav states, "The dilemma of having to talk in classical terms about phenomena which cannot be described in classical concepts is the basic paradox of quantum mechanics."[9]

The changes in thinking required to understand quantum physics do not come easily even for scientists. Zukav quotes Werner Heisenberg as saying, "Once one has experienced the desperation with which clever and conciliatory men of science react to the demand for a change in the thought pattern, one can only be amazed that such revolutions in science have actually been possible at all."[10]

A primary difference between classical and modern physics is the reluctant recognition that there is no basic building block of physical reality. There is no solid particle, however small, which forms the foundation of larger physical objects. They don't exist. Zukav states,

> "Today, particle accelerators, bubble chambers and computer printouts are giving birth to another worldview. This worldview is as different from the worldview at the beginning of this century as the Copernican worldview was from its predecessors. It calls upon us to relinquish many of our closely clutched ideas.
>
> "In this world view there is no substance.
>
> "The most common question that we can ask about an object is 'What is it made of?' That question, however, 'What is it made of?', is based upon an artificial mental structure that is much like a hall of mirrors.... The search for the ultimate stuff of the universe ends with the discovery that there *isn't any*."[11]

The implications of this one statement are enough to send classical physicists reeling. The belief in the reality of a basic building block has always been at the foundation of classical physics.

Modern physics says that instead of a solid building block at the foundational level of matter, there is only energy. Zukav states, "If there is any ultimate stuff of the universe, it is pure energy, but subatomic particles are not 'made of' energy, they *are* energy."[12] Because of the language problem, quantum physicists still make use of the term "particle" when referring to this elementary level of matter, even though "all the evidence belies the fact that quantum particles are actually particles."[13]

Nonlocality

As scientists experimented with subatomic particles, they were faced time and time again with results that made no sense. The only way for them to resolve their dilemma was to expand their minds way beyond the bounds of classical thinking. Zukav describes just such a dilemma: "A particle, as we mentally picture it (classically defined) is a thing which is confined to a region in space. It is not spread out. It is either here or it is there, but it cannot be both here *and* there at the same time."[14] In the classical view, physical objects must have some kind of "physical" contact to communicate, influence or affect one another. This could be a physical touch. It could be the sound waves created by my vocal mechanism that resonates in your eardrums and are interpreted by your brain. This could be the radio waves picked up by my radio which then produces sound waves which enter my ear and so on. In any case, the fastest that any "physical" phenomenon can travel is the speed of light, which is 186,000 miles per second—not too shabby. However, quantum theorists have discovered that subatomic particles communicate even faster. In fact, they communicate instantaneously. It is as if these particles, separated by any imaginable distance, are still in contact with each other. This characteristic is called "nonlocality." Capra states, "Whereas the hidden variables in classical physics are local mechanisms, those in quantum physics are nonlocal; they are instantaneous connections to the universe as a whole."[15]

The characteristic of nonlocality has forced scientists to speculate on the nature of the universe as they tried to come up with a theory that would make sense of their experimental results. Their conclusion is that the universe, rather than being a collection of independent parts, must be more of a whole. Capra states that "the constituents of matter and the basic phenomena involving them are all interconnected...they cannot be understood as isolated entities but only as integral parts of a unified whole."[16]

To understand this we have to get past the classical cosmology. We have to open our minds and let our imaginations develop a different way of understanding our world. No physical entities exist independently and apart from other physical entities. We are not separate from the animals or the plants or each other. Capra states, "In modern physics, the image of the universe as a machine has been replaced by that of an interconnected, dynamic whole whose parts are essentially interdependent and have to be understood as patterns of a cosmic process."[17] It may not even be appropriate to speak of "parts" of a whole. Perhaps, ultimately, "there are no parts at all in this interconnected web."[18]

Living versus Nonliving Organisms

In our classical worldview we make the distinction between living and nonliving things. We view animals and plants as living, and rocks and dirt and metal as not living. We speculate that animals and even plants may have some kind of primordial consciousness or, on the other hand, may just run on some kind of instinctual impulse. In the case of plants, perhaps "it" is just some kind of natural process or occurrence. We are certain that there are some physical objects, like rocks, that are inanimate, not alive. Well, let's take a look at what quantum physics suggests. Zukav states,

"By watching time-lapse photography we know that plants often respond to stimulae (sic) with humanlike reactions. They retreat from pain, advance toward pleasure, and even languish in the absence of affection. The only difference is that they do it at a much slower rate than we do. So much slower, in fact, that it appears to the ordinary perception that they do not react at all.

"If this is so, then how can we say with certainty that rocks, and even mountain ranges, do not react also as living organisms, but with a reaction time so slow that to catch it with time-lapse photography would require millennia between exposures! Of course, there is no way to prove this, but there is no way of disproving it either. The distinction between living and nonliving is not so easy to make."[19]

One reason for this speculation is the discovery that an object as minute as a subatomic particle may, in fact, be alive in a real and conscious sense. Zukav tells us, "The astounding discovery awaiting newcomers to physics is that the evidence gathered in the development of quantum mechanics indicates that subatomic 'particles' constantly appear to be making decisions!"[20]

Space and Time

Our customary understanding of space and time must also be reevaluated in the light of the new physics. Zukav states, "There is no single time which flows equally for all observers. There is no absolute time.... Newton made one more mistake in this regard. He said that time and space were separate. According to Einstein, time and space are not separate. Something cannot exist at some place without existing at some time, and neither can it exist at some time without existing at some place."[21] "There is no such thing as space *and* time; there is only space-time."[22]

The implications of Einstein's special theory of relativity extend to that which would seem absurd from the classical perspective. It may be that all time—past, present and future—exists now. Zukav states,

"The Newtonian view of space and time is a *dynamic* picture. Events *develop* with the passage of time. Time is one-dimensional and *moves* (forward). The past, present, and future happen in that order. The special theory of relativity, however, says that it is preferable, and more useful, to think in terms of a *static*, non-moving picture of space and time. This is the space-time continuum. In this static picture, the space-time continuum, events do not develop, they just are. If we could view our reality in a four-dimensional way, we would see that everything that now seems to unfold before us with the passing of time, already exists *in toto*, painted, as it were, on the fabric of space-time. We would see all, the past, the present, and the future with one glance. Of course, this is only a mathematical proposition (isn't it?)."[23]

After making this fantastic statement, Zukav tries to console us by saying, "Don't worry about visualizing a four-dimensional world. Physicists can't do it, either. For the moment, just assume that Einstein might be right since the evidence so far suggests that he is."[24]

Consciousness

The idea of consciousness may help bring these quantum concepts together in a way that begins to make more sense. In classical theory, consciousness may be no more than a physiological effect of the brain, perhaps just an evolutionary aberration. But quantum physics leads to the idea that consciousness may be the basic "stuff" of the universe!

Our consciousness may be a creative force. Capra thinks that scientific studies may lead to the "unprecedented possibility of being forced to include the study of human consciousness explicitly in our future theories of matter."[25] Zukav reviews the work of Carl Jung and Wolfgang Pauli (a Nobel Prize winning physicist) and states, "If these men are correct, then physics is the study of the structure of consciousness."[26] If consciousness were the basic stuff of reality, this would explain why there is no such thing as a particle or basic building block. This would explain the characteristic of nonlocality and why even the fictional particle can make decisions.

It is becoming clear, at the least, that consciousness is a creative force. Quantum physicists have discovered that the observer in a scientific experiment, merely by observing, has an effect on the actions of the observed. This effect occurs without any physical contact. There is no physical touch, no words, no sound waves, nothing but awareness or consciousness that could possibly be affecting what is observed. Capra states, "The properties of any atomic object can only be understood in terms of the object's interaction with the observer. This means that the classical ideal of an objective description of nature is no longer valid. The Cartesian partition between the I and the world, between the observer and the observed, cannot be made when dealing with atomic matter. In atomic physics, we can never speak about nature without, at the same time, speaking about ourselves."[27] Zukav states, "The new physics tells us that an observer cannot observe without altering what he sees. Observer and observed are interrelated in a real and fundamental sense. The exact nature of this interrelation is not clear, but there is a growing body of evidence that the distinction between the 'in here' and the 'out there' is illusion."[28]

We are not mere observers or bystanders in the process of the world; we are creators and integral parts of the whole. Perhaps we *are* the whole and we determine the whole. "The implications of quantum mechanics are psychedelic. Not only do we influence our reality, but, in some degree, we actually *create* it."[29] Zukav asks, "'*Did we create the particles that we are experimenting with?*' Incredible as it sounds, this is a possibility that many physicists recognize."[30]

Reality

In light of all the above perhaps we should ask, what is reality? Asking this question is a good start; it puts us in a good frame of mind—the beginner's mind—for finding the answer. Zukav states, "The world of matter is a relative world, and an illusory one; illusory not in the sense that it does not exist, but illusory in the sense that we do not see it as it really is."[31] There are many sources, both Western and Eastern, which confirm this statement. They provide many related insights that together may help us develop a clearer, more comprehensive and truer understanding of reality. Some of these sources come from academia and some are mystical in nature. There is a wide variety. It may be surprising that many principles of this new, emerging cosmology are filtering down and being applied not only in esoteric philosophies but also in practical arenas such as business and finance. The following chapters will explore some of these exciting developments.

Chapter Three
Epistemology: How Do We Know?

The following chapters will expand upon the discoveries of quantum physics and begin to apply them toward the development of a larger, more comprehensive cosmology. In this process we will use evidence not only from the "hard" sciences such as physics but also evidence from the "soft" sciences such as psychology and other academic disciplines. In addition, we will include sources that are mystical in nature. In view of this variety of sources, it would be good to begin with a consideration of the epistemological question, "How do we know?" How do we find out about the reality of our world? "Epistemology," according to Webster, is "the theory or science that investigates the origin, nature, methods, and limits of knowledge."[32] The exploration in this chapter will include an examination of different ways of knowing, including both their advantages and limitations. For many people in our culture the hard sciences are believed to be the only way we can know something for sure, so let's begin there.

The "Hard" Sciences

In the west we have come to have a great reverence for science, but this reverence has gotten to the point where many of us have forgotten that the world of science is one of hypothesis, postulates and theories more than one of facts. A good example of this was mentioned in the last chapter. The theory of the atom had become accepted as *fact* in popular culture and even among some scientists, even though it was never more than a well-accepted theory. No one had ever seen an atom, even under the most powerful microscope. Scientists did not observe its existence; they only reasoned its existence. The essence of a good theory is that it is useful for explaining whatever phenomena it is intended to explain, and the theory of the atom did this well for a long time. But as new data began to appear through new experimentation and more advanced technology, the idea of the atom had to be modified.

A theory is the best idea scientists have, given available data, for explaining why certain phenomena occur. It is a model, a theoretical construction used to explain what the world is and how it works. Theories are subjected to rigorous experimental procedures and all the known variables are strictly controlled. However, it is not possible to control the *unknown variables*. A few centuries ago scientists had no idea of the existence of the micro-biotic world. In their search for the causes of disease they were not able to factor in the presence of things such as germs and viruses. Their theories of disease and their methods of treating them were limited accordingly. A century ago scientists had no idea that consciousness could impact observed physical reality as quantum physicists have discovered. New technologies have made these discoveries possible and scientific theories adapt as new information becomes available.

The history of science is one in which one accepted idea after another has been called into question and eventually replaced by another accepted idea. We can go back to the period when the earth was obviously flat and the sun obviously rotated around it. We can go back to the period, not so long ago, when time was considered to be a constant, not relative phenomena. An example of an accepted theory that is in transition today is that of evolution. Extensive archaeological explorations over the past few decades have failed to substantiate this theory as it now stands. The evidence suggests that the theory is true for intraspecies development but not for interspecies development. In other words, archaeological evidence supports the idea that changes occur within a particular species according to the precepts of the theory of evolution, but not the idea that one species has developed from another. The missing archeological links are still missing and, with all the digging that has been done, they would have been found if they were there. We are thus in the position of being able to wonder again about how a new species first made its appearance on the earth. This wondering is the beginning of new ideas, new theories and the search for supporting evidence.

Scientific theories are limited by imagination. A hypothesis cannot be tested if it is not first imagined. It took a Copernicus to imagine that the earth might rotate around the sun and not vice versa. It took an Einstein to imagine that time and space might be interrelated, not separate. It took a Neils Bohr and a Werner Heisenberg to imagine that the world might be an interconnected whole and not a collection of independent parts. Imagination, and the subsequent discovery and perception of greater truths, is limited mainly by our conceptual, cosmological boxes. Our arrogance can keep us confined in those boxes. There is great truth in the statement that you cannot learn if you think you already know, and scientists can be as guilty of arrogance as any of us. There is also some truth in the observation, made by some, that, "science advances one death at a time." As those in power retire or die, new generations step in with new ideas and the process of imagination and discovery continues.

The problem of measurability is one that could, perhaps, sum up the limitations of traditional scientific inquiry. Science has become adept at measuring known physical phenomena, but refuses to consider the possible effect of things that it cannot measure. Science is concerned with precision and this is certainly admirable. Science wants to reduce everything to an irrefutable logical syllogism such as, "If A=B and B=C, then A=C." The problem here is the "If." Science begins all its theories with the implied, "If the physical world of the five senses is all there is, then…." Science is, to this extent, operating in an imaginary world: a vacuum or box defined as, "only that which is strictly physical and measurable." Can we truly and effectively learn about our world with this kind of limitation? Are there elements of reality that cannot be seen or measured but which, nonetheless, are part of our world and affect our experience? Quantum physicists have found that human consciousness has an effect on quantum particles, but that is merely a beginning, not an end. There may be more to our world than science can currently measure or even imagine.

The scientific method of experimentation is a valuable tool in the quest to know about our world, but to use this tool most effectively we must recognize its limitations. To be most effective, this tool must be used in conjunction with other ways of knowing.

The "Soft" Sciences

The "soft" sciences, such as psychology, are so called because they attempt to deal with phenomena that cannot be seen or measured with the exactitude of hard science phenomena. If a person is suffering from depression, for example, a psychologist can ask the person questions, make visual observations of his or her demeanor and so on, but it is not a concrete physical phenomena that can be precisely observed and measured. Depression cannot be seen under a microscope or measured with a yardstick and yet, whatever it is, depression unquestionably exists. The hard sciences do not want to deal with something as fuzzy as depression, but is it, nonetheless, something that has an effect on our physical world? If we are exploring the cause of illness, should we consider only strictly physical elements such as germs and viruses or should we attempt to determine if something like depression has an effect? If we are to truly understand our world, we cannot afford to ignore phenomena simply because they are not as precisely definable and measurable as the hard sciences would like them to be.

The soft sciences have made a valuable contribution to our knowing. A barrier to their greater success, however, is their desire for respectability. Scientific respectability is defined in our culture by the physical science model. We love the precision and irrefutable logic we would like to believe it offers. It would be great if we could have an absolute knowing about our world, but that is not currently possible. We must deal with elements and possibilities that are not precisely definable and measurable. In its desire for respectability, psychology seeks to emulate the physical science model, and in so doing it makes the same mistake that physical science does. Within the realm of its field of inquiry, traditional psychology focuses on those elements that are more definable and more measurable and ignores or berates those elements that are less definable and less measurable. It is also reluctant to consider phenomena that do not fit into the classical, Newtonian cosmology.

There is enough evidence of some kind of nonphysical reality to merit serious investigation by mainstream science. There have been scientifically oriented studies done on subjects such as ESP[33], near-death experiences[34] and even reincarnation[35]. We no longer can maintain the naive presumption that our reality is merely physical. We cannot ignore evidence simply because the subject matter is difficult to define or measure or does not fit with previous concepts.

An Introduction to Mysticism

Mysticism is not as strange an animal as it might seem to some. Mysticism should be understood as simply another way of learning about the world, which, like the sciences, has its advantages and limitations. The experience of mysticism is not limited to only a few eccentric individuals, for it is an innate ability possessed by all. Properly defined and understood, low-level mystical experiences are common among ordinary people. In our culture, however, mysticism has not been emphasized, understood or even recognized as existing. As a result, our innate abilities have been largely undeveloped and common experiences unacknowledged.

The concept of left-brain and right-brain thinking is a useful beginning for understanding what mysticism is about. In this model the left-brain operates in a linear mode; logic and language are left-brain functions. The right-brain operates in a holistic mode; it sees the whole picture, the larger context and gives us our creative and intuitive ability. We have all had the experience of logically and analytically struggling with a problem that just doesn't seem to have an answer, only to find that when we finally give up, turn our attention to something else, or just go to sleep that the solution simply comes to us. This is the right-brain at work. Writers find that thoughts often come as a block of information or a collection of interrelated ideas. This is also the right brain. The process of analyzing the ideas, fitting them in with other ideas, and putting them into the linear format of language is a left-brain function. The left-brain is a thinking, analyzing experience, and the right-brain is a sensing, feeling, perceiving experience. Intuition is a good example of a right-brain function. In intuition there comes a sense of knowing, a feeling of assurance that a certain conclusion is true, even though there has been little or no conscious analytical attention given to the subject.

Classical cosmology tells us we are isolated physical beings whose only access to the outside world is through the five physical senses. The

Epistemology: How Do We Know? 25

information we obtain about the world comes from reading, hearing, observing and the like. Many scientists and academics have stuck with this classical model even though it does not adequately account for some of the experiences in our world—even ordinary, everyday kinds of experiences. For example, someone you haven't heard from for a long time calls you on the phone just as you were thinking about calling him or her. The classical perspective is that this is simply a coincidence, a matter of odds and probabilities. If you believe in the classical cosmology, you have to come up with explanations that fit. I suggest that the "coincidences" we experience in our life far exceed their probability of occurring. The experience of intuition often involves access to information that could not be had through the normal channels of the five physical senses. A mother may sense that her child is in trouble even though the child is not physically present. People have sensed the death of a loved one before hearing anything about it. Lovers sometimes know what the other is thinking; they seem to be on the same "wavelink." The examples go on and on.

The new cosmology provides a model that helps us understand these kinds of phenomena. We no longer have to ignore them or come up with improbable explanations. The new cosmology suggests that the universe exists more as an interconnected web, at the foundations of which are things like energy and consciousness rather than the solid physical structure of the old cosmology. Quantum particles have been found to be "aware" of each other at any imaginable distance and, if quantum particles can do it, we should not consider it so strange that humans can. The discoveries of quantum physics thus provide a context in which the experience of mysticism at all levels can be better understood. Mysticism, basically defined, is simply the ability we each have to access information beyond our immediate physical surroundings. Mystical experiences do not have to be religious in nature. Mysticism is simply the ability we have to be aware of information that could not be had through the five physical senses.

The right-brain is thought by some to operate as a receptor of that which is beyond our physical self and not just as an isolated processing unit. In the new cosmology creative insights may be the interaction of our intelligence with thoughts, ideas and consciousness beyond us. In the old cosmology our creative ideas come from electrical impulses shooting around within a mass of tissue. In either case, the key to accessing the right-brain is to temporarily suspend the activity of the left-brain. We can do this by simply relaxing: stop struggling, stop thinking and analyzing, stop concentrating. When we are in a state of relaxed peace and good feeling, undistracted, we can most effectively access the right brain. It is said that Einstein got some of his best ideas while relaxing in the bathtub. Eastern religions emphasize the need for meditation, which is essentially no more than the suspension or relaxation of left-brain activity. Performed correctly, meditation is supposed to lead to mystical experience and insight. Women in our culture are traditionally associated more with intuition and men with logical analysis but, with practice, anyone can learn to easily alternate between right- and left-brain functions. In our culture logical analysis has been more highly valued, but it is time to explore, understand and equally value the right brain.

The Limitations of Mysticism

The high-level mystical experience works something like this: the mystic receives what could be called a block of thought, a perception, or a sense of knowing. The format of this experience could be a vision, the hearing of words, or simply a conceptual knowing. In any case, what is perceived must go through an unconscious translation process before it reaches the conscious mind of the mystic. The mystical experience is thus an *interaction* between some reality out there and the mind of the individual. That which is perceived by a mystic at the conscious level is limited by and determined by the ability, background and concepts of the particular mystic. The mystical experience is an *interpretation* of reality, *not* the reality itself; a mystic cannot see or perceive that for which they have no concept.

The fundamental process of mystical vision is not different from ordinary vision. In ordinary vision, sensory input in the form of light waves makes contact with receptors in our physical eyes, is interpreted by our brains, and the result is what we see. We do not, technically speaking, directly "see" an object. There is always this filter or interpretation that comes between the object and our perception of it, and this process produces some fascinating effects. When the sensory input is incomplete our minds sometimes "fill in the blanks" and what we see is a whole picture. Optical illusions make us see what is not really there. These illusions depend on preconceived notions to fool us and we cannot truly see what is there until our concept of what we are looking at changes. In drawings that contain hidden pictures we may not see that which is right in front of our eyes until we have an *idea* of what to look for. What we see in the physical world is, nevertheless, relatively consistent because the physical world is familiar to us. We have developed similar expectations and concepts of what is out there. In mystical vision there is no such familiarity and much less consistency in regard to expectation and concept. Mystical vision could thus be expected to be less reliable than ordinary vision.

The hearing of words involves a similar process of interpretation. We are all aware of how easy it is to misunderstand someone and how what we hear can be influenced by our background, expectations and concepts. A humorous example is the many ways in which children will mangle the words to the song, "A Star Spangled Banner," which, as we all know, begins, "Jose' can you see?"

But there are other issues in regard to words. Words merely are symbolic representations of a thought or perception. If I want to express a concept, I choose the word-symbols I believe represent my thought. The hearer must interpret those symbols and associate them with a concept that is in their own mind. Words can thus be an inept means of communication, particularly when the speaker is trying to express a concept with which the hearer is not familiar. As human beings we have developed words to represent what we see and what we think. Obviously we have not developed words for that which we have not yet conceived. If a mystic reveals to us the thoughts of God or of some higher intelligence or some reality beyond our normal experience, can we expect these things to be adequately expressed in human language? I like the way Trappist monk Father Thomas Keating addresses this issue. He says, "Silence is the language God speaks, and everything else is a bad translation."[36]

Every mystical "revelation," old or new, must be understood in this context. Every written or spoken revelation and every vision has come through a human filter. This is not to deny the ability we have to access information in this manner, but it is to acknowledge our limitations in perceiving and communicating it. Many Christians would like to believe that the words of Jesus are the pure, unfiltered thought of God. But even if it were so, we must still acknowledge the limitations of words to capture the mind of God. Further, the words of Jesus were intended for a specific audience at a specific time and place. When Jesus spoke of Satan or Hell, we should not assume that he believed in the reality of those things. He was trying to reach people who lived in a certain place and time and had to express his truths within the confines of their language and concepts.

Summation

We are entering a new era in our understanding of the world, propelled and validated by the discoveries of quantum physics. Our other academic disciplines must get up to speed and be willing to think outside the mechanistic box of classical cosmology. To meet this challenge we must utilize both our left-brain and right-brain capacity and be integrated thinkers. We must sense, feel and intuit that which is true and then logically analyze and experiment on that which is perceived. In this process it is valid to consider the thoughts, perceptions and conclusions of mystical sources as well as scientific and academic sources.

Chapter Four
The New Cosmology

The new cosmology involves an expanded understanding of the nature of the world, how it works, and who we are as human beings. Perhaps the basic difference between the old cosmology and the new cosmology is the realization that the world is not composed of solid, physical matter and does not, therefore, have the limitations that would be inherent in that kind of substance. Instead, the basic "stuff" of the world is energy or consciousness, or perhaps "conscious energy." It is difficult to define. Because of this foundation, the world does not exist as a collection of independent, isolated parts, but exists more as an interconnected whole. It is referred to as a "whole" because its parts can have instantaneous communication, awareness and influence on each other. Larry Dossey, M.D., in his book, *Recovering the Soul: A Scientific and Spiritual Search*, spends many pages detailing evidence and scientific proofs of the nonlocal nature of the world in general and of human consciousness in particular. He says, "An invisible connectivity unites all things no matter how disparate."[37]

This world is not the mechanistic place of predetermined "natural" occurrences presented in the classical model; it is a dynamic and responsive world. As mentioned in chapter two, physicists have discovered that the micro world of quantum particles is responsive to human thought, but the evidence also extends into the macro world. The use of visualization techniques by athletes is a common example. In a visualization procedure athletes will imagine exactly what they want to do and how they want to do it, recognizing that this deliberate imagining helps to bring about the desired result. It is recognized that the *belief* that one will win or be successful helps to bring about that experience, just as the belief in failure or fear of failure helps to bring about failure. Here, for example, is a story about Arnold Palmer:

> "Golfer Arnold Palmer has never flaunted his success. Although he has won hundreds of trophies and awards, the only trophy in his office is a battered little cup that he got for his first professional win at the Canadian Open in 1955.
>
> "In addition to the cup, he has a lone framed plaque on the wall. The plaque tells you why he has been successful on and off the golf course. It reads:
>
> If you think you are beaten, you are.
> If you think you dare not, you don't.
> If you like to win but think you can't,
> It's almost certain that you won't.
>
> Life's battles don't always go
> To the stronger woman or man,
> But sooner or later, those who win
> Are those who think they can."[38]

In the past we thought our beliefs were merely a reflection of the reality of the world, but now we must ask, what is that reality? If there is something I want to accomplish, is the reality that I can do it or that I can't do it? And how is the eventual outcome influenced by what I believe? On an interpersonal level we notice that our experience with a person, positive or negative, is influenced by our beliefs about them. On a developmental level, one child may be taught that they are bad, another taught that they are good, and each grows up to act accordingly. Was the eventual reality set at birth or was it influenced by what the child came to believe about themselves? We could go on and on with examples but clearly the "reality" we experience is a malleable event, influenced by what we believe. We have tended to think that our beliefs about self, others or the world reflect the reality of our world. To a great extent it is actually the other way around. The reality of our world reflects our beliefs.

The effectiveness of visualization and belief does not, in itself, indicate the need for a new cosmological perspective, for they can be explained within the classical framework. We can say, for example, that visualization by an athlete creates a "muscle memory" pattern in the brain that the body then follows during physical performance. The belief that one will win can be said to be effective because it causes the athlete to be more focused, less distracted by the fear of losing. These explanations are not necessarily correct, but they are reasonable within the classical context. They make sense within that framework.

Other evidence of the connection between thought and physical experience stretches the boundaries of the classical cosmology a little more, although it can still be made to fit. There is, for example, increasing evidence of the link between our thought and our physical health or illness.[39] One possible explanation for this is that our thoughts somehow affect the functioning of our immune system, making us more or less susceptible to disease and illness. Positive thoughts improve the functioning of the immune system and negative thoughts reduce it.

Evidence that stretches the classical boundaries even further is provided by Jacob Liberman, O.D., Ph.D., in his book, *Take Off Your*

Glasses And See. In this book Liberman states, "Almost every day I hear of a person who has dramatically improved his or her visual acuity by changing his or her way of thinking...I know of many people who have experienced spontaneous, permanent visual improvement simply by realizing that it was possible, and releasing their limiting beliefs."[40] At this point people stuck in the classical cosmology will tend to ignore the evidence or simply say they don't believe it. Still, perhaps even this could somehow be explained within the classical ideology.

The evidence gets really interesting when we begin considering the effect of our conscious thought on the world outside our physical bodies. Here we begin to go beyond the domain of current mainstream science and academia because here we must begin to envision a new cosmology. Here we must begin to visualize the energetic, interconnected world of the quantum physicists. It has been well documented, for example, that human thought can affect the health of plants.[41] Even though this evidence is obtained by scientifically oriented experiments in which the variables are controlled and so forth, much of mainstream science and academia will tend to ignore it, *simply because it does not fit into their cosmological framework!* Rather than ignoring the evidence, why don't we reexamine the cosmology that says it can't be true?

The evidence is mounting that our conscious thought can influence the *events* of our lives. It influences whom we come in contact with, the "coincidences," circumstances and conditions that we experience. The old cosmology attributes these things to random, impersonal, chance occurrences. But the influence of thought, belief and imagination is now being recognized in many arenas, including the very practical one of business, finance and wealth. The positive thinking "movement" so popular in the field of sales is a reflection of this understanding. Dr. Tag Powell in his book, *Think Wealthy*, states, "Part or all of your financial problems may be the fact that you don't *think* wealthy. Your first level of change will be, to paraphrase Rene Descartes, 'I think, therefore I am—I think wealthy, therefore I am.'"[42]

There is also increasing evidence that the kind of conscious thought we call "prayer" has a real effect on our physical experiences. There are numerous studies to this effect, but I quote here from a recent front-page article in the *Kansas City Star* newspaper titled, "Power of prayer gets scientific boost":

> "More than faith, it may be fact: Ill patients fare better when people pray for them, even when they're unaware that they're the subject of prayers, and even when they don't know the people doing the praying.
>
> "So says a study conducted by researchers at Kansas City's Mid-America Heart Institute at St. Luke's Hospital and published today in *The Archives of Internal Medicine*. The study, one of the largest of its kind, offers the most compelling evidence yet linking improved health to what is known as off-site or remote intercessory prayer."[43]

We do not need to assume this phenomenon is the result of intercession by a divine being. There are too many theological problems involved with believing that God acts in response to prayer, and I will address these theological issues later in the book. The effectiveness of prayer may, in fact, be further evidence of the power of our conscious thought to influence the world around us. If a study were conducted with those who believed their conscious thought could influence the health of another, and who directed their positive thought accordingly, the results may well be the same as it is with prayer.

The explanations we develop for observed phenomena derive from our assumptions about reality. It is important that we allow our assumptions and explanations to change as the demands of new evidence make the need for this change apparent.

Consciousness

Quantum physicists have stated that it may be more accurate to say that a quantum particle *is* energy, rather than *made of* energy. This energy appears capable of making decisions and is responsive to human thought, so perhaps we should refer to it as "conscious energy" or as "responsive consciousness" or perhaps just "consciousness." But what is consciousness? Is it just some bizarre evolutionary aberration of physical matter? The evidence is clearly indicating that consciousness neither resides in the physical brain nor is caused by the physical brain. It is something that is "nonphysical," something that exists outside of a purely physical understanding of the world. Biologist Rupert Sheldrake believes the physical brain acts more as a receptor of information and conscious thought, than as the source of it. Sheldrake suggests that our thoughts, desires, intentions, memories and so forth, do not physically reside in the brain, and he uses the analogy of a television set to explain his position:

> "A search inside your TV for traces of programs you watched last week would be doomed to failure, for the simple reason that your set tunes in to TV transmissions but does not store them.... Damage to the set might destroy certain channels or diminish the clarity of the reception, yet this is not evidence that the sounds and pictures you see originate inside the TV set itself."[44]

Rupert Sheldrake is no rogue philosopher. He received his Ph.D. in biochemistry and cell biology from Cambridge University and has been the recipient of a number of prestigious fellowships. Sheldrake has recognized the inadequacy of mechanistic explanations of the world we live in. He says these explanations are "essentially an act of faith in the mechanistic method, not really a strict scientific hypothesis."[45] We can consider, for example, the mechanistic explanation that DNA is the controlling force responsible for the development of physical organisms. This position is held despite the fact that "Bone and ear and liver cells all contain the same DNA, [and] so there must be something over and above the DNA itself to account for why they turn out differently."[46] Sheldrake comments on this problem as follows:

> "DNA, by providing the code for the sequence of amino acids, enables the cell to make particular proteins. That is *all* DNA can do.... But the problem with morphogenesis is not just the question of getting the right proteins in the right cells at the right time. It's how, given those proteins, the cells organize themselves into particular forms; how cells group together in tissue of particular forms; and how those are shaped into organisms of particular forms. DNA helps us to understand how you get the proteins which provide, as it were, the bricks and the mortar with which the organism is built, but it doesn't explain how these bricks and mortar assemble into particular patterns and shapes. The idea of DNA shaping the organism or programming its behavior is a quite illegitimate extrapolation from anything we know about what DNA does.... Everything to do with heredity and properties of living organisms is being projected on to DNA within the mechanistic model—all the unsolved problems of biology are being attributed to DNA.... So what starts as a rigorous and well-defined theory about the way DNA codes the RNA

and how RNA codes the proteins, soon turns into a kind of mystical theory in which DNA has unexplained powers and properties which can't be specified in exact molecular terms in any way at all."[47]

Sheldrake has developed his own theory to explain how and why physical organisms develop as they do. His theory of "formative causation" is consistent with the new cosmology being developed by quantum physicists and others. In this theory Sheldrake proposes the existence of what he calls "morphogenetic fields," which are essentially a form of consciousness. These fields are a sort of nonphysical pattern that guides the form and characteristics of each particular physical species. These patterns are not static, for each generation adds what it has learned back into the pattern, changing it for future generations. Support for this theory is provided by studies of rats and other animals in which it has been demonstrated that later generations learn a task with ever increasing speed, *even when there is no genetic connection!*[48] In other words, animals in different parts of the world, with no genetic connection to the trained animals, learn the task quicker than preceding generations!

A Mystical Perspective

To this point in the chapter we have been reviewing the evidence gathered from the hard and soft sciences. There is related evidence and thought from mystical sources. We would do well to listen to the voice of mystics and test their conclusions as much as possible. We should look to see if the perceptions of mystics are consistent with scientific evidence, but we should not value one way of knowing over the other. Mystics may, after all, reveal truths that science has not yet discovered.

Abraham-Hicks is a mystical source that provides a relevant perspective at this point in the discussion.[49] The world of Abraham-Hicks is the same energy-based, connected universe that we have been considering in the previous pages. Abraham-Hicks also recognize the effect our thoughts can have on our physical experience, but use the word "vibration" because it has a greater depth of meaning. Our vibration consists of our desires, intentions, thoughts, beliefs and feelings. We constantly emit this vibration and the universe responds to it in the same way that quantum particles respond to the thoughts of quantum physicists. We do not really know why this response occurs, just as we do not really know why a physical mass has a gravitational force. With gravity, we just say it is a physical law. Abraham-Hicks say the basic law of the universe is the "Law of Attraction," which states, "That which is like unto itself is drawn."[50] In other words, vibrational forces are drawn to each other and *everything* has a vibrational force. Thoughts have a vibration, physical objects have a vibration, people have vibrations. As conscious, aware, intelligent beings, we have the ability to *decide* what our vibration will be and other vibrational forces in the universe are attracted to it. Let's take a closer look at how this process works.

A thought is a vibrational force and it attracts like thoughts. As we focus on a particular subject, we get more and more thoughts on the subject. When we focus on a desire we begin receiving ideas of how to get what we want. But our vibrations also attract physical experiences. If we focus on our desire for a new car, for example, we start becoming more aware of advertisements or we begin encountering more people of like mind and interest. If we focus on a particular kind of car, we start seeing them everywhere. The ideas, advertisements, people and cars were always out there, but they were drawn into our personal experience by our vibrational force.

We are complicated beings and our vibrations can also get complicated. If we believe we can't afford the new car or that it is impractical, the resulting vibration counteracts the desire and reduces its effectiveness. The effectiveness of a vibration depends on its strength. We can have strong beliefs and desires, and we can have weak beliefs and desires. We can have conflicting beliefs and desires. Stronger and purer vibrations have more attracting power and result in stronger, clearer response from the universe.

Abraham–Hicks say that our emotions indicate the strength of our vibration. A strong, pure desire will result in strong positive emotion. But a strong desire that is *contradicted* by strong belief results in a strong *negative* emotion. If we really want something but believe we can't have it, we experience anger or depression. The desire to be loved and appreciated is often a powerful one. When we believe we are loved we feel wonderful and when we believe we are not, we can be angry or devastated. Weaker desires and beliefs result in correspondingly milder emotions. Abraham-Hicks talk about this in terms of energy. They say a powerful desire summons powerful energy from the universe, and when we allow that energy to flow through us we feel exhilarated. According to Abraham-Hicks, the energy of the universe is only positive; there is no negative energy. But when we summon this energy by our desire and then block it by our beliefs, the pent-up energy results in a feeling of anger or frustration.

Abraham-Hicks say that desire is our lifeblood for it summons the life-giving energy of the universe to us. People with no desire or who consistently block the energy get sick or die. The Law of Attraction is how Abraham-Hicks explain the evidence reviewed earlier in this chapter. The Law of Attraction explains why an athlete who believes that he or she will win, attracts that experience. It explains why a person who thinks wealthy would attract wealth. It explains why a person who thinks sick gets sick and why a person who thinks healthy stays healthy.

As we seek to realize the true nature of the world in which we live, we have to look at the evidence and the explanations as a whole, determine which theory most adequately explains all the available evidence, and then experiment more with that theory. As discussed previously, the classical cosmology cannot adequately account for much of the evidence that has been accumulating in recent years. Theories such as the "Law of Attraction" merit further consideration.

Chapter Five
A Nonphysical World

It has become clear we cannot continue thinking of our world in strictly physical terms. There is something going on behind the scenes. Our fundamental reality is nothing like we thought it was; it is bigger, more dynamic and more malleable than we ever thought possible. Our role in it is more important, more central, and more powerful than we had imagined. We are not so much bystanders, subject to outside forces as we are creators. To appreciate this more fully we need to take a deeper look at who we are and at the larger context of our experience.

According to Abraham-Hicks, the world we live in is primarily nonphysical. This nonphysical aspect is an ongoing, eternal reality that is larger than and prior to the physical world. The physical world, however, is not separate from the nonphysical; it is part of that larger reality. This is perhaps more understandable when we realize that physical matter is not a different substance from nonphysical energy; matter is essentially a condensed and focused form of the same energy. The world of our physical experience did not occur by mere chance. It was intended, designed and brought into being by our nonphysical consciousness.

Human beings are essentially nonphysical beings. We were nonphysical before we were born into the physical world, we are primarily nonphysical while we live in the physical world, and we will still be nonphysical when we die. We cannot be separate from our nonphysical self; it is who we really are. The great Christian mystic Teilhard de Chardin also perceived this truth. He said, "We are not human beings having a spiritual experience. We are spiritual beings having a human experience."[51] Our consciousness, which includes our beliefs, intents, desires, knowledge and thoughts, is *who we are* in an essential and eternal sense; it is the nonphysical part that does not die when the body does. Our nonphysical self, however, is much larger than simply our current conscious thoughts. It is the older, wiser, broader part of our self, for it has learned from all of our many previous life experiences. According to Abraham-Hicks, we have had many past lives and will continue to be born into the physical realm for as long as we desire. All that we have learned, thought and experienced in our previous lives is fed back into and retained in the nonphysical self; it is a growing, changing, dynamic consciousness. We can observe a correspondence between Abraham-Hicks' description of our nonphysical self and the theory of morphogenetic fields discussed in the previous chapter. In each case the accumulated learning, retained in the nonphysical part, influences the subsequent physical experience. The knowledge retained in the nonphysical allows us to more quickly adapt to the circumstances of a new physical life experience. Our previous life experiences also greatly impact the intentions and desires we bring with us. A person who was poor in a previous life may enter the next life with a desire for material wealth. A person who was famous may desire relative anonymity to preserve a sense of privacy. A person may come with a desire to teach or help others. The desires we bring are general in nature and as we encounter the specific circumstances into which we are born, we decide how best to actualize them in that context. Our desires may not be fully realized in a given lifetime, but we are happy when we are in touch with these desires, believe we can accomplish them and work toward that end.

A question that often arises in regard to reincarnation is, "Why don't I remember my previous lives?" Abraham-Hicks state this is because of a conscious decision at the nonphysical level for, if we remembered all the details, we would have little room in our minds for new experience. Each new physical life is intended to be a fresh new experience, unencumbered by memories of hurts, fears and previously held beliefs. As adults we sometimes find ourselves burdened by memories and constricted in our development by the perceptions and beliefs that have become hardened in our minds. A saying that relates to this is, "You can't teach an old dog new tricks." A new physical life is intended to be a fresh start. The specific memories of previous lives are not normally available, but we are still influenced by our larger nonphysical self as described above.

Religion

Abraham-Hicks do not use religious language in their communication of the world they perceive, but acknowledge there have been many enlightened beings on our planet such as Jesus and Buddha and many others we have never heard about. "Enlightenment" is, of course, a relative term indicating one's ability to perceive the true nature of reality. In this context we may even consider quantum physicists as enlightened to some extent. In any case, when listening to prophets or scientists we must remember that what they offer, at best, is a *relative truth* given in a certain context and influenced by certain assumptions. Abraham-Hicks do not use the term "God" much, but when asked about God's existence I have heard them say, "That is all we talk about."

Most religions, to one extent or another, emphasize ideas of guilt, punishment by God, sin and the like. The Eastern concept of "karma" is one of these. Abraham-Hicks offer a different picture of the ultimate, eternal reality. They say we are not here to prove anything to God nor to anyone else. We are merely here, by choice, to experience the wonders and joy and learning that are possible in a physical environment. There is no condemnation, no judgment, and no punishment in the hereafter. The idea of guilt is strictly a human invention. Abraham-Hicks refer to the nonphysical realm as a place of pure, positive energy: a place of uncompromised well-being. They say this well-being also permeates the physical plane even though many people do not recognize it. Abraham-Hicks desire to teach us to recognize it now but, in any case, when we die we remember once again the unconditional love or pure, positive energy that is the source of all things. Many people cannot conceive of a temporal or eternal world without guilt. They cannot conceive of a world in which bad people are not required to pay for their sins. Our religious beliefs and concepts fundamentally derive from and are limited by our cosmological ones. Let's examine this relationship a little closer. We can begin with the basic questions of life and death. In classical cosmology we are thought to be simply bodies. All consciousness and "beingness" is contained in that physical organism and when the body dies, the person simply ceases to exist. In this cosmology it would certainly be horrible to have one's brief life ended prematurely or scarred by some negative experience. If we can imagine, however, a cosmology in which we come from a pure, positive environment, enter into the physical world as often as we like, and then return to that positive environment, we would have a totally different view of physical life experiences. We would not take them so seriously. We would be easier, more relaxed about our experience here. We would, for one thing, be more willing to take the chance of pursuing what we really want. We would be less afraid of failure. We would not be afraid of death. This exercise in imagination does not, obviously, prove anything about the true nature of reality. From it, however, we can begin to get an idea of how our religious beliefs reflect the cosmology in which we choose to believe. As we consider different concepts of reality, Abraham-Hicks advise us to trust our emotions. They suggest we ask ourselves, "Which feels better?" for according to Abraham-Hicks, our emotions guide us toward truth.

In classical cosmology the universe is a mechanistic and impersonal place. I say "impersonal" because we have relatively little control over what happens to us. Illness is largely a matter of genetics and germs. Accidents and violence are largely the result of chance occurrences. We can educate ourselves about the world and take precautions but so much of our life's experience is still beyond our control. This lack of personal control is, in fact, a *defining characteristic* of a mechanistic cosmology. This is what the term "mechanistic" essentially implies. From this cosmological context we thus seek the cause of our experience *outside* ourselves, and this leads to a culture of victimhood. We blame genetic inheritance, other people, God and psychological scars caused by parents for our life experience.

The theological belief in guilt and punishment finds support in this cosmology, but was really founded in a more ancient cosmology. The world was once thought to be controlled by God or by the gods. Personal control was a matter of figuring out how to appease this God or gods, and the theories varied greatly. The theories have ranged from baptism to animal sacrifice to rituals of all kinds, ancient and modern. In an important sense this previous cosmology is not so different from the classical cosmology, for both attribute our life's experience primarily to *something outside of us*! In one case it is the gods, and in the other it is "nature." The new cosmology is not impersonal and does not lead to a belief in the idea of guilt.

The world of classical cosmology is a physical world of physical activity. In it we believe that physical "doing" is the only way to change the world in which we live. But our furious activity often seems to be little more than a finger in the dike. When we solve one problem another comes up to take its place. The world of the new cosmology is primarily a nonphysical place in which our thought has a much greater effect than our physical actions. We must come to recognize that our thoughts, not our actions, are the primary *effective* level of our existence. Psychologists Frances Vaughan and Roger Walsh state, "You may believe that you are responsible for what you do, but not for what you think. The truth is that you are responsible for what you think, because it is only at this level that you can exercise choice. What you do comes from what you think."[52] As a society we need to seriously consider some alternative beliefs. We need to consider a new cosmology.

There Is No Law of Assertion

The physical and nonphysical are not two separate realities; they are an *interactive whole*. The Law of Attraction is a way of describing how they interact. We have reviewed evidence of how our thoughts and beliefs—our vibration—affect our bodies, the events of our lives, and the world around us. Much of this evidence is already common knowledge. People recognize that if they believe in themselves and their ability to succeed, they probably will. Many people have observed that when they expect good things to happen, beneficial coincidences and opportunities seem to come to them. In the field of medicine we find that, "For the first time in history more Americans went to alternative health care providers last year then to allopathic physicians."[53] The transition in our cosmological perspective is thus already under way. However, in this transitional period we find that people will give themselves or God credit for the good things that happen, but still tend to blame others or God for the bad things. People have not yet fully recognized that their fears, negative thoughts and beliefs also attract conditions and events into their lives. It doesn't matter if our thoughts and beliefs are negative or positive, they still influence this dynamic, malleable world of energy in which we live. If we feel from somewhere deep inside that we deserve punishment, the Law of Attraction states that someone or something will sooner or later appear to accommodate us in this regard. People are accident prone not because the usual "odds" somehow do not apply to them, but because they have come to think of themselves that way; because they have come to believe it is so.

An important converse corollary to the Law of Attraction is that there is no "Law of Assertion."[54] In other words, while the Law of Attraction states that conditions and events of like vibration are attracted into our experience, the converse is also true; conditions and events that do not match our vibration *cannot* enter our experience. In real-life terms what this means is that if we are not afraid of crime or violence and don't focus on them, they will not enter our experience. According to the vibrational laws enunciated by Abraham-Hicks, a person who wants to commit a criminal act will attract a victim whose thoughts and fears are a match to that vibration. This idea is clearly in direct conflict with the impersonal perspective of classical cosmology. This particular conflict may be a difficult hurdle for many people as we continue to move toward new understandings of the world. In our culture many people may actually *like* the idea of being victims because they then have someone or something to blame beside themselves for their misfortunes. But the cosmology that Abraham-Hicks present has nothing to do with guilt. In their view, negative experience is merely an opportunity to reflect on the thoughts that attracted it, and to consciously intend what we would prefer. The physical world is like a playground or learning experience that has no effect on the pure, positive energy of the eternal, nonphysical world. In their view, we are all eternally safe.

Some people may have trouble believing we create our negative experience because they do not believe they have these kinds of thoughts. In our culture we are inundated with reports of crime, disease, accidents and other tragedies. We live in a negative society. The negativism is so pervasive that it is considered normal to complain, criticize or be concerned about something. Counteracting this takes a deliberate intention to consciously decide what we will focus our thoughts on, and few people have developed this skill to any great extent. Most of us simply "tune-in" to the negative vibrations that surround us, but this negativity does not truly represent the reality of our world. According to Abraham-Hicks, our planet is inundated with well-being. The fact that a car accident is news, for example, reflects the fact that the vast majority of people get to where they're going without incident! We decide whether to focus on the negative or the positive. Our feelings and experience will reflect our decision.

Time and Testing

According to Abraham-Hicks, a difference between the nonphysical and physical realms of experience is that of time. There is no time in the nonphysical realm. In that realm our thoughts will immediately attract the corresponding experience. In the physical realm, however, there is usually a time lapse between thought and experience. The length of the time interval is determined by the strength and purity of the vibration being offered. A strong, clear desire or fear will manifest more quickly. Conflicting thoughts or vague desires and fears take longer or may never manifest in physical form. This time lapse can make it difficult for us to see the connection between our thought and experience.

The connection between thought and feeling is more immediate and obvious. If we have a strong desire for something, and believe we can have it or accomplish it, we will experience enthusiasm or excitement and the experience comes fairly quickly. If we don't believe we can have what we desire, the energy will be blocked and we experience a negative emotion. The strength of the negative emotion depends on the strength of the energy summoned. If our beliefs block a mild desire with its mild energy, we may experience frustration. If we block a strong desire, we may experience anger. The old cosmology tended to blame people for being angry, but Abraham-Hicks give us a way to understand what is really going on. To avoid strong, negative emotion we could simply reduce our desire; if we don't care what happens, we are not going to react strongly. But according to Abraham-Hicks, desire is our lifeblood. Through desire we summon the life-giving, positive energy of the universe. In the view of Abraham-Hicks, it is better to change our thoughts and beliefs than to reduce our desire. This can be difficult at first but it just takes practice. This practice will not only make us feel better but will also help create the experiences we want in our lives.

Our emotions are not *caused* by what happens to us, what other people do, what they say about us, and so on. In every situation it is solely our own thoughts that determine how we feel. If we drop a glass in the kitchen and break it, we feel a negative emotion if we think ourselves stupid or think about how we don't want to clean it up. But, we don't have to think those things. If someone says something negative about us, this can trigger our own negative self-thought and result in negative emotion. If we maintain our positive self-image (self-thought) it will not bother us. Losing our jobs creates significant negative emotion when we start imagining ourselves running out of money, losing the house, and so forth. Positive emotion results when we imagine ourselves finding a great new job, one we like even better than the last.

The variables of time, intensity and purity of desires, fears and beliefs can make it difficult to objectively test the theories of Abraham-Hicks. But if we wanted to, we would find a way. With experience we could develop increasingly more refined and accurate methods for testing. The theories of Abraham-Hicks hold more promise for understanding the world of our experience than the classical cosmology. It is already clear that we feel better with positive thought and modern psychology is aware of this. Psychologist Shelly Taylor states, "Positive illusions about the self, the world, and the future may not merely promote mental health, but may actually be essential to it."[55] But people are not satisfied with merely living their lives as an "illusion." The unique contribution made by the new cosmology is the possibility that our positive thought may actually have a real effect on our physical experience.

Influence

Each person is like a universe unto himself or herself. The stream of nonphysical energy is so limitless, and the possibilities of physical experience so varied, that each person's life is like a world of its own, a world which they themselves have created. For each individual thought a matching experience is waiting to happen. We can experience riches or we can experience crime, depending upon our thoughts, beliefs, desires and fears. It is all out there. However, thoughts and beliefs have no power to control the experience of another. We cannot cause the experience of another. Nevertheless, because most people have not yet learned to consistently choose their own vibration, the vibration of whoever or whatever is around them will often be an influence. Thus a particularly positive and happy person at work or at a family gathering can lift everyone else up, while a particularly negative person can bring others down. This is not a matter of control but of influence. The influence of others begins in childhood as we take on the beliefs and attitudes of our parents, other authority figures, and our peers in regard to the world and in regard to ourselves. These beliefs then become a determining influence on what we experience in our lives. As adults we read the newspaper, listen to the evening news and continue to take on the vibration of what we are observing, never recognizing the power of our thought. But if we only knew, if we were only taught, about the power of belief, imagination and thought, we could become deliberate creators of our own life experience.

Chapter Six
A New Theology

We can observe some strong correspondence between the perceptions of mystics and prophets and the cosmology currently being developed by scientists and academics. There is, however, a language problem, for many mystics tend to express their insights in religious terminology. If we can look past that, we will see the correspondence. I use the term "mystic" in a general sense to describe those people who purport to have a *primary* knowledge of a reality beyond the five physical senses. Jesus, for example, would fall into this category as I define it. A theologian is a person involved in the *secondary* process of analyzing and interpreting the teachings of mystics, such as those contained in scripture. Theologians interpret the "enlightened" teachings of mystics from the context of their own assumptions, cosmological and otherwise. This has often resulted in bizarre religious doctrine. This chapter will focus on theological issues from the perspective of the new cosmology.

As discussed previously, quantum physicists state that the universe exists as a unified whole in which energy is the basic component of all things. There is no fundamental separation of parts in this cosmology as there was in the classical cosmology. Mystics have also perceived this wholeness and unity. If, for the moment, we could think of God as a sort of universal positive energy, we will see the similarity between that which mystics and scientists are saying. Saint Paul stated, "There is one God and Father of us all, who is above all and through all and in all" (Ephesians, 4:6). Joseph Smith, Jr. expressed a similar perception when he stated that God is "in all things, and is through all things, and is round about all things: and all things are by him, and of him; even God, for ever and ever."[56] The Apostle Paul also expresses the idea in another way: "For I am sure that neither death, nor life, nor angels, nor principalities, nor things present, nor things to come, nor powers, nor height, nor depth, nor anything else in all creation, will be able to separate us from the love of God in Christ Jesus our Lord" (Romans, 8: 38-39). St. Augustine in his book *The Confessions* says, "'But Thou wert more inward to me, than my most inward part; and higher than my highest.'"[57] Jesus said, "I am in my Father, and you in me, and I in you" (John, 14:20).

Whether we call it a "universal conscious energy" or "God," the underlying perception is the same. This positive energy permeates all things and is the fundamental reality of our existence. The remaining question is whether all things *are* God, or whether God exists as a separate being who is merely "in and through" all creation. Frankly, nobody knows the answer to this question, but we do not need to resolve it to continue making discoveries about our world. What is becoming clear is that there is an intelligence and order to the world in which we live. Further, there is a "love" or positive energy that is the source and foundation of all this.

Emphasis in this chapter will be on Christianity in general, and the teachings of Jesus in particular. When we revisit the recorded teachings of Jesus in light of the new cosmology, we can observe a strong correspondence with that cosmology. It is almost as if two thousand years ago Jesus perceived the reality we are just beginning to understand today. Jesus did not use modern words such as "consciousness," "positive energy" or "nonphysical;" he was not speaking to modern persons.

Jesus presented his viewpoint within the confines of the language and concepts of his time. Nevertheless, the basic truths he presented are similar to those of the "new" cosmology.

Ask and Ye Shall Receive

Jesus understood the creative power of our thought and desire when he said, "Ask, and it will be given you; seek, and you will find; knock, and it will be opened to you. For *every one* [italics mine] who asks receives, and he who seeks finds, and to him who knocks it will be opened" (Matthew, 7:7,8). Traditional Christianity has, of course, interpreted this to mean that God is the one who responds to us and grants what we ask. The new theology (based on the new cosmology) suggests that God has set up the universe in such a way that the "universe" responds to us. Traditional theology takes the viewpoint that humans are relatively weak and powerless beings dependent on God to do things for them. The new theology says that we are powerful creators of our own experience. This new perspective is expressed well in the book *Conversations with God*:

"If you believe that God is the creator and decider of all things in your life, you are mistaken.

"God is the observer, not the creator. And God stands ready to assist you in living your life, but not in the way you might expect.

"It is not God's function to create, or uncreate, the circumstances or conditions of your life. God created you, in the image and likeness of God. You have created the rest, through the power God has given you. God created the process of life and life itself as you know it. Yet God gave you free choice, to do with life as you will."[58]

Traditional Christianity has had to deal with the fact that people often do not get what they ask God for. One explanation offered for this is that sometimes the answer is "no." Another explanation is that we are not good enough or not obedient enough to receive God's blessings. There are many other explanations offered even though Jesus clearly said, "*Every one* who asks receives." The new theology states that we *always* get what we ask for; the universe responds to *every one* of us, *all the time*. The new theology also defines what "asking" is. Asking is our true thoughts and beliefs—our vibration—rather than our words, for our words do not necessarily correspond to what we are really thinking or feeling. We have all experienced a situation in which somebody said something like, "It's nice to see you" or "I feel fine" and you could tell it wasn't what they really meant. What the universe of energy hears from us is our vibration, not our words. If someone asks God to heal him or her of some illness, but at a deeper level they really believe it's not going to happen, the universe, responding to their vibration says, in effect, "Okay, it's not going to happen." In the language of the universe, the level of our true thoughts and feelings, they have gotten exactly what they asked for.

Jesus identified a key element in this process of asking or, in the new terminology, "creating." This element is "faith." In traditional Christianity, faith is basically interpreted as a relatively pure belief that God will respond to our supplications. If we have doubt, God will not respond. This, minus the religious aspect, is precisely the position of the new cosmologists. Our positive or negative expectations—our faith or doubt—have a creative effect on our experience. But the extraordinary truth of the matter, from the new perspective, is that it *doesn't matter* what we believe in! We can believe in ourselves, we can believe in God, or we can believe in fate. We can believe in genetics, we can believe in a pill the doctor gives us, or we can believe in a religious ritual. But it is our *belief* that is the effective agent in this energetic, responsive universe. *We* are the creative agents in this physical playground and, in fact, we could do much more than we currently believe. The only limitation to our experience is our own belief about what is possible. Jesus said, "For truly, I say to you, if you have faith as a grain of mustard seed, you will say to this mountain, 'Move from here to there,' and it will move; and nothing will be impossible to you" (Matthew, 17:20). It is certainly worth noting here that Jesus did not say *God* would move the mountain for us; he clearly implied that it would be done by the power of our word alone.

But faith is not simply a cold, intellectual belief; it involves emotion. Theologian Dr. Robert Mesle says that faith is the sensation of being "grasped" by something that is of great value to us.[59] As discussed previously, Abraham-Hicks also talk about the integral importance of emotion. If there is something we really want, and we truly believe we can have it, we will experience a very strong, positive emotion. This is faith, and it is a creative phenomena.

Resist Not Evil

Jesus also seems to be aware of the power of negative thought. He gives us the following admonition: "But I say to you, do not resist one who is evil. But if any one strikes you on the right cheek, turn to him the other also" (Matthew, 5:39). It is not healthful, either physically or emotionally, to harbor resentments toward another person or to entertain negative thoughts of any kind. "Resisting evil" serves to bring the negative elements of the world more prominently into our thoughts. At a feeling level, our focus upon it increases our fear and sense of vulnerability. That kind of focus, as discussed previously, only serves to bring more of what we don't want into one's experience. Not resisting or "turning the other cheek" can be understood as not giving our attention to the negative aspects of the world. *A Course in Miracles* expresses the idea of "resist not evil" as follows: "In my defenselessness my safety lies."[60] If you *knew* you were safe, what would be the point in defending yourself? I use the word "know" here to describe faith: that coming together of thought and feeling. There is a tremendous variety of experience possible in this world; the focus of our attention will determine which part of that variety we will personally experience.

Society in general is so far from understanding these great insights that most simply go on with their usual struggle against the evil forces of the world. Doesn't it seem clear from our experience that the more we, as a society, focus on crime and fight against it the worse it gets? Or the more we focus on the problems of drugs, cancer or AIDS, the worse they get? It may seem incredible and contrary to "common knowledge" that fighting against things only increases them, but this is what the evidence is beginning to indicate. Social programs that work are the ones with a positive orientation; the ones that give a person hope for the future and a positive image of themselves. A person with an attitude and experience of poverty will not be able to climb out of that world until they can *imagine* a different experience and believe in that vision. Treating our prisoners like "dirt" *will not* rehabilitate them, but treating them with respect may give them a new conception of the world and of themselves.

The antithesis of resisting evil is love. At the basic level, "love" is a positive focus and resisting evil is a negative focus. Jesus emphatically admonished us to focus on love. He said we should love our enemies (Matthew, 5:44). He said, "You shall love the Lord your God with all your heart, and with all your soul, and with all your mind. This is the great and first commandment. And a second is like it, You shall love your neighbor as yourself. On these two commandments depend all the law and the prophets" (Matthew, 22:37-40). But love is not primarily an action; it is a thought. Perhaps we could best describe this thought as one of appreciation or gratefulness. When we appreciate the predominant well-being on our planet, we feel love. When we appreciate the good aspects of another person, while ignoring their bad aspects, we feel love. *Love is any thought that brings the positive energy of the universe into our experience!*

Love, Guilt and Free Will

Traditional Christianity teaches that God is love, but also teaches the concept of guilt. In fact, love and guilt cannot coexist; they are mutually exclusive. We can verify this easily from our own experience. In any moment when you are feeling love, there is no fear or guilt in regard to yourself or anyone else. But in the next moment, if your thoughts change and you begin feeling fear or guilt, there is not love. If God is love, there is no room for guilt in God's mind.

Traditional Christian theology teaches many things that do not make much sense. It teaches that we have "free will," but also states that we will be punished if we do not act in accordance with what God wants us to do. The book *Conversations with God* comments on this as follows: "There are those who say that I have given you free will, yet these same people claim that if you do not obey Me, I will send you to hell. What kind of free will is that? Does this not make a mockery of God—to say nothing of any sort of true relationship between us?"[61]

The idea of free will makes a lot more sense in the new theology. From this perspective there is nothing we as human beings can do that will alter the love or positive energy that is the eternal foundation of all existence. Neither do we have the power to *control* the experience of another or assert experience onto them (see Chapter Five). In the new theology free will only means we have the power to *personally* experience whatever we choose. We can choose to experience guilt and the feeling of separation from God. We can choose to believe horrible things or good things about ourselves and others, and our actions and experience will flow from these beliefs. But whatever we choose, there is no condemnation from God. And why would there be if the only thing we can effect is our own temporal experience?

To understand this better we might consider a number of analogies. Our experience in this world has been compared to that of a dream in which, no matter how good or bad the dream, when we wake up the real world has not changed. We find that the dream has not affected our real and eternal self. *A Course in Miracles* states, "When you wake you will see the truth around you and in you."[62] Another way of looking at this life is that God simply sent us out to play, and it doesn't matter if we make mistakes, get into arguments or lose whatever game we are playing. When we go back home we are still safe and warm and loved. In *Conversations with God*, "God" states,

> "I do not care what you do, and that is hard for you to hear. Yet do you care what your children do when you send them out to play? Is it a matter of consequence to you whether they play tag, or hide and seek, or pretend? No, it is not, because you know they are perfectly safe. You have placed them in an environment which you consider friendly and very okay."[63]

Jesus also uses an analogy to help us understand our eternally safe condition. In the parable of the prodigal son, the son leaves home, spends all his money and gets himself into an awful circumstance. Yet when he returns home he is welcomed with open arms and is given a feast (Luke, 15:11-24). Our doubts about our eternal safety and happiness are responsible for the creation of our greatest enemy: fear. The old theologies are often based upon fear and serve to perpetuate it.

There are many other aspects of theological and religious doctrine that need to be reconsidered in light of the new cosmology, but the discussion in this chapter has covered the basic ones. Love is the most basic issue, but it needs to be understood properly. The next chapter will be devoted to a more detailed consideration of this foundational concept.

Chapter Seven
Love and Meaning

The energy at the foundation of all things is a life giving force. When we allow that energy to flow through us we feel good. When we experience that energy we feel happy, inspired, we feel…love. Love is what we feel when this energy is flowing through us, for it is a positive energy. There is no negative energy. As human beings we can experience the energy flowing or we can experience its absence, just as we can turn the light switch on or off. There is no "dark switch" that can be turned on because darkness is simply the absence of light. We turn the switch on by thinking thoughts that allow the energy to flow. What are the thoughts that bring this power into your experience? As a general rule, positive beliefs about ourselves, others, or the world allow this energy, but the specific thoughts vary with each individual.

As a definition of love, then, we could say that love is this positive energy that is in and through all things and which is the foundation of all things physical and nonphysical. We could thus say that love is all there is. On the other hand, we could say that love is the thoughts that allow the energy to flow. It really doesn't matter how we define love as long as we understand the basic principles involved. One thing we can be sure of is that love is not an action. We can perform a seemingly loving act with no love in our hearts and we can sit at home by ourselves and have a powerful love flowing through us. It is our thoughts, not our actions that determine and control the flow of this energy.

We have referred to this positive life force as "love," but we could use other words. Like a rose, it is delightful no matter what we call it. In fact, it might be helpful to use other words because there are so many misconceptions about what love is. When the universal energy is freely flowing through us, with no resistance on our part, we feel joy, and this is synonymous with love. When we have a positive belief and expectation for something we really want, we feel excitement. This is the same energy. It doesn't matter what the object of our love, joy or excitement is. A heartfelt statement such as "I love ice cream," or "I love my car," or "I love a warm spring day" allows that positive life force to flow through us. This flowing of energy is clearly evident to anyone who has ever felt good about any of these things. It is sometimes thought that the test of our love is the degree to which we are involved with helping others or involved in activities for the common good. This is not correct. Love is *who we are* and each of us is entitled to experience that feeling of knowing who we are regardless of anything we might or might not do for others in this physical life. The "test" of our love is simply the way we feel: the extent to which this positive energy is flowing through us.

Helping Others

The experience of love in any individual life is *always* helpful to others, whether it is specifically intended to be or not. This is because of the influence we have on one another. When we encounter someone who is connected to this source, excited about their life, involved in activities they like and letting the positive feelings flow, we are influenced in that direction. When we encounter someone who is worried, frustrated, and otherwise unhappy, we are likewise influenced. We could thus say, paradoxically, that in order to help others we must be selfish. We must do what we need to do, think what we need to think to be connected to this source before we can be of benefit to others.

Some of us have a specific desire to help others. As with any desire, we should focus on this, nurture it, and let it grow within us. If our beliefs do not contradict this desire with thoughts such as, "There is nothing I can do," we feel good as we imagine ourselves in helping situations. As we focus on the desire we begin to get ideas of how we can help, and we find that ways and means of service open to us.

Actions, in and of themselves, are not neither helpful nor destructive. When we do something for someone else, does this influence him or her to feel worthy and capable, or unworthy and incapable? Teaching a person to fish is obviously preferable to giving them a fish, but more important than what we do is what we think. To be truly helpful to another we must respect them; we must see them in a positive light, see only their good qualities. We must have positive expectations for them and positive beliefs about the world of our mutual experience. This will influence the other accordingly, and nothing is more effective than changing a person's mind. We will know our thoughts are in alignment with our purpose by the way we feel. When there is joy in our hearts, we can be assured that the actions which flow from us will be helpful.

The Nonmoralistic Nature of Love

A fundamental characteristic of the universe is that "It" or "God" or whatever term we want to use, is not moralistic or judgmental in any way. We can see this quite clearly by simply looking at the world around us. We can observe, as the author of the New Testament book of *Matthew* has, that "God makes his sun rise on the evil and on the good,

Love and Meaning 63

and sends rain on the just and on the unjust" (Matthew, 5:45). We can observe that people whom we consider unjust or uncaring are sometimes healthy and successful while those we deem good may experience illness and financial struggle. Of course, the reverse is also often true: "good" people are sometimes successful while "bad" people are not. In sports we observe that some highly successful, wealthy athletes are outstanding citizens while others don't give a damn about anybody but themselves. There is no pattern of success or failure, good fortune or tragedy, that relates to one's moral character.

Christians generally will acknowledge these facts, mainly because they are indisputable, but state that justice, in the form of reward and punishment, is meted out in the afterlife. This is not so. In the previous chapter we discussed the logical incompatibility of love and guilt. We observed that God could not love *and* see guilt, for the two states of mind are mutually exclusive. Jesus tried to get at this all-encompassing nature of love when he stated that we should forgive a person seventy times seven times (Matthew 18:22). But the main argument against this idea of eternal or afterlife punishment is that it is completely unnecessary. As human beings we cannot change the fundamental, eternal reality of existence. Neither can we *control* the experience of another human being. So, if the only thing we can affect is our own temporal experience, be it positive or negative, what need would there be for punishment?

This question of the moral perspective of the universe can be explored a bit further. The world of our physical experience is intended (by our larger, nonphysical self) to be a place of learning, discovery, experimentation, and enjoyment. This is clear from the wide variety of sources from which people derive the feeling of success, enjoyment and meaning. The universe does not care whether we want to race model airplanes, find a cure for disease, be a religious leader, start a business or spend our time alone in the mountains. The universe supports whatever we want! The life giving, positive energy can flow to us regardless of our specific desire or object of attention. The only sin, if there were such a thing, would be in not choosing, for it is our desires that summon this life force. The universe does not care about the specific object of the desire; it simply wants us to experience and learn from whatever we choose.

Allowing

Love is all-inclusive. The energy will not flow to us when we think negative thoughts about any person or thing. The more positive thought we have for the objects of our attention, the more constant will be our experience of love. We do not, however, have to take on the impossible task of having positive thought toward every minute aspect of the world. In fact, if we had only one object of love and spent all of our time focused on it, we would feel love constantly. Those who have a particularly strong love for a grandparent, parent, child, or intimate partner, find that a focus on this person will carry them through the day in a positive feeling state.

Most of us, however, cannot maintain a constant focus on the people and things we love. We have certain negative memories that keep popping up, or certain fears, prejudices and other negative thoughts. Maybe certain people in our experience really annoy us. It is certainly appropriate to ignore or forget about these things when possible, for it is not our responsibility to change the world or change other people and, in fact, we cannot do it. We can only control our own experience. But if we cannot ignore or turn our attention away from some particular person, situation or memory, we must transform our thoughts on the subject. In the case of a person we don't like, for example, we choose to see only their good points. When we do this we find the feelings of love beginning to flow through us again, and we find our experience with this person changing for the better.

Our beliefs can get in the way of changing our minds in this manner. If we believe we are in eternal danger and need to justify ourselves, or if we believe that people and situations can assert themselves into our experience, we will necessarily be fearful. These basic beliefs bring fear into our experience, and from fear come guilt, condemnation, hatred, anger and prejudice. From fear we begin to divide the world into those people and things we consider safe and those we do not. But with love there is no fear. Love connects us rather than separates us. Love allows other people to be different from us, hold different beliefs, different desires, intentions, and so forth. Love does not resist or fight against these differences. It is more difficult to allow when we have traditional religious beliefs of sin and punishment, or when we have a classical cosmological perspective, but love is its own teacher. As we begin to allow, we begin to feel positive energy flowing through us and it feels good. We begin to recognize which thoughts allow that energy to flow and which do not, and we begin to have less tolerance for the negative emotion to which we had become accustomed. We can thus be led into an understanding of love by love itself. Nevertheless, it is helpful to begin with a set of assumptions that are conducive to this process.

Meaning

The experience of meaning is something we all want in our lives. It is something that some people spend their lives searching for and the lack of which makes life, well…meaningless. But, given the foregoing, meaning is really not that hard to understand or even to find. The experience of meaning is really nothing more than the experience of this positive energy in our lives. To experience a sense of meaning, we simply acknowledge the principles outlined in this book and *choose the thought that feels better!* Our emotions guide us into these simple but profound truths.

Chapter Eight
Medicine and Psychology

The concepts of the new cosmology have important implications for virtually every field of endeavor, including such a diverse range as sports, finance and the criminal justice system. In this chapter we will focus on some of the ways these new understandings can revolutionize the practice of medicine and psychology. In fact, the revolution has already begun. The ideas presented here are rapidly becoming more and more mainstream; we increasingly see them being presented in mainstream publications as more and more supportive studies come forth.

Medicine and Health

Some people might consider this statement bold: Doctors do not really understand what causes disease, illness and other unhealthful conditions, nor do they know how to cure them. The most that medical professionals currently can do is simply observe the correspondences between disease and cure, and draw conclusions or develop theories based on these empirical observations. The word "empirical" is a holy one among scientists, but empirical studies only reveal *what* happens, not *why*, and if you don't know the "why" of it, the "what" is open to a wide margin of error. To use a barbaric example, bloodletting was formerly used as a means of curing illness. Physicians used this method because it was observed that patients *sometimes* got better after being subjected to this treatment. Doctors didn't know why it seemed to help, although they had their theories. Some of today's medical practices, such as chemotherapy, are equally barbaric. Despite their sophisticated sounding explanations, doctors do not really know why chemotherapy sometimes works and sometimes does not. Nearly every medication, treatment or preventative measure used today is inconsistent in its effectiveness. Recent studies, for example, have shown that a low-fat, high-fiber diet *does not* reduce the risk of colon cancer.[64] This was shocking to the medical profession because the idea that it did had been well accepted and well documented by previous empirical studies.

The reason scientists and medical professionals cannot understand the "why" of disease is because they have been locked into an inadequate cosmological box. The well-known author, Bernie Siegel, M.D., states, "Scientists and physicians often have tunnel vision because of their training."[65] They have viewed people as physical objects under the control of natural laws only. From this perspective they were unable to see the larger forces at work; they could not understand the dynamics of energy and the power of human thought and belief. Today, however, "scientists have been doing a tremendous amount of research linking consciousness, psychosocial factors, attitudinal healing and immune function."[66] They are finding that "Love, hope, joy and peace of mind have physiological consequences, just as depression and despair do."[67]

One source of evidence of the power of thought and belief comes from the well-known "placebo" effect, through which a person develops positive expectations of wellness from taking an inert substance. The "nocebo" effect is when a person develops negative expectations (such as side effects) from taking the substance. Siegel states, "With both placebos and nocebos, it is the *expectations aroused* by the substance or procedure that are ultimately responsible for the result. Sometimes the effect can be induced simply by the words or attitude of a doctor or other authority figure."[68] Siegel tells us that, "Generally speaking, about one-third or more of the people treated with placebos report positive results."[69] He cites this example:

> "We have cut his prednisone dosage in half as he was really getting nasty mood swings. To restore his hair growth we rubbed a 'magic mixture' on his head and told him it would make his hair grow. It did! When we stopped using it, it quit growing, and started growing again when we resumed putting it on.
>
> "When Kelly is on prednisone he eats like a horse and when he is off he has almost no appetite at all. To help out his suffering appetite I have been giving him folic acid out of the bottle for his prednisone, which he calls his hungry pills. Lo and behold his appetite has returned via the placebo prednisone."[70]

Additional evidence of the power of the mind comes from the study of people with Multiple Personality Disorder (MPD). Siegel discusses this evidence:

> "There are certain physiological traits that we assume to be fixed, like diabetes, left- or right-handedness, allergies or color-blindness. It appears, however, that people with MPD may be allergic to cats or orange juice in one personality but not in another, may exhibit burns in one personality but not another, may show drug sensitivities in one personality but not another, may switch from being right-handed to being left-handed. I knew

someone who had to keep half a dozen different pairs of glasses in her bedside stand, because she didn't know who she would be when she woke up."[71]

Siegel sums it up by saying, "What's in your mind is often quite literally, or 'anatomically,' what is in your body."[72]

Doctors rarely have an explanation for why some people recover from incurable diseases. Siegel suggests we get to know these people better and even *ask them* what they think. He relates his research as follows:

> "It's incredible to think of all these thousands of people who recovered from 'incurable' illnesses and were never asked how or why they thought they had gotten well. When you do ask, as I have done and as researchers more receptive to this kind of thinking have also, you find that over 90 percent of the people will tell you about a significant change in their life prior to the healing. An existential shift has occurred in them, and for the first time in their lives they are truly living. They don't see their disease as a sentence but a new beginning."[73]

The desire to live and the conscious awareness of something to live for are key ingredients in recovery from illness, as well as survival in general. The central point in Victor Frankl's monumental book, *Man's Search for Meaning*, is that the survivors of the Nazi concentration camps were the ones who experienced a powerful *reason* to keep on living.[74] Recovery and survival are not a matter of odds; they are a matter of the power of human thought and desire! Desire summons the energy of life, the life force, and keeps us alive. No power in hell, no circumstance or "natural" cause, no Nazi or disease, can overcome the individual power of thought and desire. This is the world God has created for us.

In learning to tap into this power over our own destiny, the main point we must keep in mind is that our feelings are the best indicator we have of our dominant thoughts and beliefs. As discussed previously, if our thoughts and beliefs are allowing that positive, healing, life-giving energy of the universe to flow through us, we feel good! The more we allow it, the more joyous and loving we feel. Most people who are suffering from some so-called incurable disease may desire to live, but they don't experience the hope, faith and happiness of anticipated recovery because their belief system will not allow it. Medical professionals and other authority figures often add to the problem because of their own negative expectations and beliefs.

But how can people experience love, joy and peace when faced with the daily pain of some debilitating disease? There may be as many individual answers to this question as there are people, but some general suggestions apply to all. "Many techniques for achieving peace of mind are available. These include hypnotic suggestion, biofeedback, relaxation training, visualization, yoga and other consciousness-altering techniques."[75] Siegel tells us, "Relaxation is so commonly acknowledged to be effective that some hospitals now broadcast relaxation programs on closed circuit television in the patients' rooms. The list of diseases altered in a positive way by relaxation would fill this page."[76] Simply imagine the outcome you desire and let your feelings be your guide. Focus on your reasons for living and let these fill your mind. Select the thoughts that feel good for, in truth, "happiness is a choice. Its source is within you."[77]

Scientists place far too much importance on statistics as a means to "prove" causation. They say that people who have a certain gene or eat certain foods are more likely to get certain diseases. They can't explain why some people get them and some don't, they merely say it is a matter of odds. We do not live in a random universe. The new cosmology suggests that statistics do not prove that one physical thing sometimes causes another; they merely show how the majority of people are thinking! Siegel states, "The future is not determined by statistics; it is determined by individuals."[78]

These days we are given so much advice on what to eat and what not to eat, what to do and not to do, and the advice keeps changing. People make themselves miserable trying to follow these prescriptions to live longer. What's the point? While our food and physical activities do have some effect on our well-being, the most important factor, far and away, is the joyous energy we allow into our souls. Siegel suggests that to have a long, healthy existence we need to celebrate life. He says, "Let there be no need for therapy in heaven to work out resentments of all the things you did to not die, and all the fun you missed out on while exercising, meditating and preparing your vegetables."[79]

The new cosmology does not deny the existence of germs, viruses and the like. These elements are around all of us all the time. The main question is why these elements make some of us sick and not others. The old cosmology states that it is because of impersonal, chance occurrences. The new cosmology states that it is because of the power of our individual attraction. There is an extremely wide variety of elements and circumstances in this physical realm; our personal vibration determines what we attract.

Psychology

The new cosmology also has significant implications for the practice of psychological counseling and therapy. If we reflect back on the way our thoughts affect the way we feel and the way they affect our experience in the world, we would have to seriously question some of the methods used in traditional psychological practice. Psychology is a broad field, containing a wide variety of approaches to therapy, but many of these approaches tend to focus on the problem rather than the solution. They focus on a person's negative experience, fears and what they don't like about their life, rather than on what they *want* to experience. As should be clear from preceding chapters, a focus on what we don't like about ourselves, other people or the world serves to make us feel bad, and serves to *perpetuate* those things in our lives! To be who we want to be and experience what we want to experience we must focus our thoughts on what we want, not the opposite.

There is a place in therapy for the examination of painful memories, but they must be used correctly. The contrast of experience we observe around us or internally in our own lives is actually a good thing. It is because of this diversity that we can begin to identify what we want. The things we do not like in our lives or in the world clarify what we do want; the more intense the negative observation, the clearer the desire for something else. Our desire summons the life-giving energy, and without the contrasting experiences of our world, we would not know what we wanted. If red was the only color in the world, we would not have the enjoyment of picking out the color of our cars, houses or clothing. The contrasting experiences in our lives can serve this clarifying purpose if we will just look at them without dwelling in misery, and simply decide what we prefer. Our negative experience can effectively be used in this manner.

Another mistake often made by traditional psychology is the belief that something that happened in the past has *caused* us to be who we are today. This is the same fallacy of correspondence that we observed in the field of medicine. There is, without a doubt, a correspondence between childhood abuse and adult feelings of insecurity, lack of confidence, or depression, but the *causative factor* is the thoughts of unworthiness, guilt and fear that the child develops under those circumstances, not the event itself. It is vitally important to understand the true cause, because this will determine the appropriate treatment. It is not necessary to relive those painful events, get angry with those responsible, complain, or otherwise continue to dwell on them; in fact, this can be harmful. A person does not need to go back and resolve all those old issues; instead, they need to change the thoughts they are thinking *now*! It may not be easy to change a habitual pattern of thought, but at least it is effective.

In most cases it is not necessary to review or address past events or issues at all. Most people already know how they want to feel and what they want to experience, and these positive images should be the focus of therapy. Past issues and events can be effectively and productively and simply *ignored*. However, in some cases the issues a person has are so powerful that they cannot simply ignore them; the feelings and memories keep coming up despite their best efforts to focus on something more positive. In these cases it is necessary to specifically address one's feelings about a particular person or set of people, and we must transform our view of these people. We must focus on their good points and on what we like about them. If necessary we must give them excuses such as, "They did the best they knew how." By transforming our thoughts about the person we will either find our relationship with them transformed, or that we will be able to let them go. We will know we have succeeded by the way we feel.

Today the field of psychology is opening to the perspectives I have been presenting. The sub-discipline of "Positive Psychology" is gaining momentum and receiving mainstream recognition. Martin E.P. Seligman, Ph.D., former president of the American Psychological Association is one of the main proponents and has been called "The Freud of the next century." Another major proponent is Mihaly Csikszentmihalyi, Ph.D., author of the national bestseller, *Flow: The Psychology of Optimal Experience*.[80] The mainstream magazine, *Psychology Today*, has openly come out in support of this movement. Robert Epstein, editor-in-chief, states, "A research-based initiative that focuses on 'joy,' 'optimism' and 'happiness'—on human strengths rather than human failings – is a fad that's long overdue and that just might last. Is my bias showing?"[81] *Psychology Today* discusses the work of Seligman and Csikszentmihalyi as follows:

"Seligman believes that only a small number of the 18 million people diagnosed with depression actually suffer from biologically based depression, which, he says, means our conception of depression is all wrong. It is not something created by rejection or childhood traumas that make us feel bad or say negative things, he says. It's much less complex than that. Maybe, 'what looks like a symptom of depression – negative thinking – is itself the disease,' Seligman says. This thought has driven him to devote a large part of his life's work to learning how to change patterns of negative thinking.

"Mihaly Csikszentmihalyi, Ph.D., is Seligman's partner in the positive psychology movement. As a young boy in Hungary, he witnessed firsthand the devastation of World War II, leaving him, ironically, with a sense of awe: How do we explain those people who remain strong during war, those who are able to withstand tragedy, who are able to live a happy life even when everything else is falling apart? he wondered. Young Csikszentmihalyi's fascination drew him decades later to positive psychology research.

"Csikszentmihalyi, a professor at Claremont Graduate University, co-authored with Seligman the introduction to the *American Psychologist's* January 2000 special issue on happiness, excellence and optimal human functioning."[82]

Recent research and thought promises a bright future for the field of psychology and for those in need of what it has to offer.

Chapter Nine
Personal Application and Conclusion

Despite the broad implications, the principles of the new cosmology are, at the heart, very personal. In truth, they can only be applied one person at a time, for it is our individual thought that is the predominant power and influence in the world of our experience. In this chapter we will discuss how an individual can apply these principles in his or her own life and suggest some methods for so doing.

In Western culture we have not been trained to choose the thoughts we prefer, so it takes a bit of practice. In Western culture we have, instead, been taught that action is the means to accomplish whatever it is we are wanting. But action is hard work when our thoughts are not aligned with our intended purpose and often ends in failure under those conditions. If it is our desire to start a business of our own but our thoughts reflect ideas such as, "I don't deserve it," "I can't do it," or "Most new businesses fail," our chances of success are indeed small. On the other hand, if our thoughts are aligned with our desire, we find that our actions are much more pleasurable and productive. We find that things just seem to fall into place and that "coincidences" and inspirations line up to support our purpose.

As we set about learning to consciously select our thoughts, we might, at first, attempt to begin monitoring all of them, only to discover that *this is not possible*! Nor is it necessary. We have a fantastic number and variety of thoughts that pass through our heads every day, and it is certainly not necessary or possible to be consciously aware of each one. Instead, we merely have to monitor our feelings. Our emotions will inform us of the *balance* of our thought. The most important factor in creating what we want in our lives (if you remember nothing else, don't forget this) is that we feel happy! When we feel good we know the energy of the universe is flowing into us and creating what we want in our lives. So, if you are not feeling good, *choose a thought that feels better!* If you find that you cannot feel good about some subject of importance to you, think about something else! If necessary, distract yourself by going to a movie, taking a nap, or whatever helps you get into a better feeling place. *Do not*, under any circumstances, continue to push against the problem or sit wallowing in your misery. This is counter-productive.

Methods

The best method to use in learning to consciously determine your own thought is simply the one that works for you. Accessing this personal power is not a matter of finding and using the right technique; it is a dynamic, individual process. Nevertheless, in this section we will look at some suggested methodologies that might be helpful.

Writing our thoughts down is a good way to help us focus on them, so one method is to write about what we want to experience. We can imagine and write about our desired experiences in enough detail that it seems we can taste them *right now*! This process is both creative and enjoyable. Another method is to make a list of affirmations that we will review or repeat to ourselves every day. Using affirmations can be helpful, although if they are too specific, we may find that they don't feel quite true, as if we don't really believe them. And, in fact, that is the case. In this situation, our ingrained beliefs are contradictory to what we would like to believe and it is best to begin with more general kinds of affirmations. If we want to buy a new car but don't believe we can afford it, we can begin with general statements such as, "People buy new cars all the time," or "I am a powerful attractor of whatever I want." We simply look for statements that feel good to us. Once we are comfortable with the general affirmations we can begin getting more specific. The key to these methods is in paying attention to how the writing or affirmations make us feel.

Imagination and visualization are powerful tools. Time spent in positive imaging and visualization are never wasted; they are actually more productive than action in regard to the ratio between time spent and results obtained. Role-playing is a method that can be effective for some people. Another method is prayer. But remember, our prayers must be prayers of thankfulness for what has happened and for what is to come. Prayers of complaining do not bring us what we want.

We have been talking about positive thought, but meditation, practiced correctly, is the experience of "no thought." The experience of no thought is a peaceful, relaxing, positive state of being. This is because love is our natural state, and when we don't offer any thoughts that prohibit that energy, we feel it. Meditation can be an effective method for getting ourselves into a good feeling place, and an important element of our overall practice. The feeling of no thought is certainly better than the feeling of negative thought, but the feeling of positive thought is better still. Positive thought is exhilarating; no thought is peaceful and relaxed.

No matter which method or combination of methods we use, the main point is to keep ourselves in a good feeling place. If we find ourselves in a negative frame of mind and do not seem able to change our thoughts, the best method is to distract ourselves. Perhaps the best advice is simply to enjoy life. Choose the thoughts that make you feel good and the things you want will come your way. It is the law of the universe.

Conclusion

We began this book with a discussion of what a cosmology is and how it can both help and hinder our attempts to understand the world we live in. A cosmology can become a self-validating conceptual box when we see it as fact and not merely assumption. The classical cosmology has served us well for some time, but the evidence no longer supports it. Quantum physics is leading us into new conceptions of reality and paving the way for a more accurate and comprehensive understanding of our experience. As we journey into these new understandings we must reevaluate the question of how we can best learn about the world. Science has value but also limitations. Academic inquiry likewise has value and limitation. Mystical inquiry, not well understood in our culture, also has value and limitation. Our best chance of knowing comes from integrating all these approaches.

The evidence is leading us to see the world as one of energy and connection, not solid matter and separation. In this world we find that thought is more powerful than action. We find ourselves creators of our experience, not victims of outside forces. A great variety of scientific and academic sources are contributing evidence to this new vision of the world. Some mystical sources are providing clarity and direction to our search.

The implications for theology are quite astounding. The fields of medicine and psychology are being dramatically impacted as well. Writers and researchers in these disciplines, as well as many others, are transforming our thinking.

In the past we have ignored or discounted evidence because it was not consistent with our cosmological assumptions. We now have a model that includes and makes sense of all the available evidence. This model is a practical one that empowers the individual and helps us understand our experience. The new cosmology gives us a positive and exciting vision of the future of humankind.

About the Author

David grew up near New York City but has spent most of his adult life in the Midwest. He is a former Christian minister who has spent many years studying Christianity, Eastern religion, New Age philosophy, psychology and science. David holds two master's degrees and possesses a unique ability to bring divergent perspectives together to find the truth that underlies them all. He expresses his understandings in clear, concise, accessible language.

You may contact the author at:
1008 NW Redwood Circle
Blue Springs, Missouri 64015

or at:
adjudd@msn.com

Notes

1. *A Course in Miracles* (Glen Ellen, CA: The Foundation for Inner Peace, 1975), Workbook, p. 451.
2. *The Interpreter's Bible, Vol. 1* (New York: Abingdon Press, 1952), p. 462.
3. Fritjof Capra, *The Tao of Physics* (Boston: Shambhala Publications, 1991), p. 53.
4. Ibid., p. 62.
5. Gary Zukav, *The Dancing Wu Li Masters: An Overview of the New Physics* (New York: Bantam Books, 1979), pp. 107-108.
6. Ibid., p. 117.
7. Ibid., p. 118.
8. *Webster's New Universal Unabridged Dictionary* (New York: Simon and Schuster, 1983).
9. Zukav, p. 202.
10. Ibid., p. 192.
11. Ibid., pp. 192-193.
12. Ibid., p. 193.
13. Ibid., p. 47.
14. Ibid., p. 47.
15. Capra, p. 310.

16. Ibid., p. 309.
17. Ibid., p. 330.
18. Ibid., p. 329.
19. Zukav, pp. 45-46.
20. Ibid., p. 47.
21. Ibid., p. 148.
22. Ibid., p. 149.
23. Ibid., p. 150.
24. Ibid., p. 150.
25. Capra, p. 319.
26. Zukav, p. 31.
27. Capra, pp. 68-69.
28. Zukav, p. 92.
29. Ibid., p. 28.
30. Ibid., p. 28.
31. Ibid., p. 155.
32. *Webster's New Universal Unabridged Dictionary* (New York, Simon & Schuster, 1983).
33. Jean Millay, Ph.D., *Multidimensional Mind* (North Atlantic Books, 1999), et. al.
34. See the works of Raymond Moody, M.D., Melvin Morse, M.D., et. al.
35. Brian Weiss, M.D., *Many Lives, Many Masters* (New York, Simon & Schuster, 1988). Tom Shroder, *Old Souls* (New York, Simon & Schuster, 1999).
36. Larry Dossey, M.D., *Healing Words: The Power of Prayer and the Practice of Medicine* (San Francisco: Harper, 1993), p. 87.

37. Larry Dossey, M.D., *Recovering The Soul: A Scientific and Spiritual Search* (New York: Bantam Books, 1989), p. 179.
38. "Bits & Pieces", Vol. R, No. 33 (Fairfield, NJ: The Economics Press), pp. 11-12.
39. See the works of Larry Dossey, M.D., Bernie Siegel, M.D., et. al.
40. Jacob Liberman, O.D., Ph.D., *Take Off Your Glasses And See* (New York: Crown Publishers, 1995), pp. 12-13.
41. Peter Tompkins and Christopher Bird, *The Secret Life of Plants* (New York: Avon Books, 1973).
42. Tag Powell, Ph.D., *Think Wealthy: Put Your Money Where Your Mind Is!* (Largo, FL: Top of the Mountain Publishing, 1991), p. 18.
43. Eric Adler, "Power of prayer gets scientific boost" (The Kansas City Star, 10/25/99).
44. Dossey, pp. 27-28.
45. Ibid., p. 197.
46. Ibid., pp. 197-198.
47. Ibid., pp. 198-199.
48. Ibid., pp. 190 ff.
49. Tapes and books are available from Abraham-Hicks Publications, San Antonio, TX.
50. Jerry and Esther Hicks, *A New Beginning II* (San Antonio: Abraham-Hicks Publications, 1996), p. 102.
51. Stephen Covey, *The 7 Habits of Highly Effective People* (New York, Simon & Schuster, 1989), p. 319.
52. Frances Vaughan, Ph.D. and Roger Walsh, M.D., Ph.D. *Accept this Gift* (New York: Perigee Books, 1983), p. 26.
53. "Conversations", No. 38 (Ashland, OR: ReCreation), p. 8.

54. Jerry and Esther Hicks, *A New Beginning II* (San Antonio: Abraham-Hicks Publications, 1996), p. 102.
55. Shelley Taylor, Ph.D. *Positive Illusions: Creative Self-Deception and the Healthy Mind* (New York: Basic Books, Inc., 1989), p. 226.
56. *Doctrine and Covenants* (Independence, MO: Herald Publishing House, 1986), 85:10c.
57. Louann Stahl, *A Most Surprising Song: Exploring the Mystical Experience* (Unity Village, MO: Unity Books, 1992), pp. 87-88.
58. N.D. Walsch, *Conversations with God: an uncommon dialogue, book I* (New York: G.P. Putnam's Sons, 1995), p. 13.
59. C. Robert Mesle, Ph.D., *Fire in my bones: A Study in Faith and Belief*, (Independence, MO: Herald House, 1984), p. 34 ff.
60. *A Course in Miracles* (Glen Ellen, CA: The foundation for Inner Peace, 1975), Workbook, p. 284.
61. Walsch, p. 39.
62. *A Course in Miracles*, Text, p. 102.
63. Walsch, p. 13.
64. Gina Kolata (New York Times), "Cancer Data on Fiber Diet Stun Science" (The Kansas City Star, 4/20/00).
65. Bernie S. Siegel, M.D., *Peace, Love & Healing* (New York: HarperPerennial, 1990), p. 3.
66. Ibid., p. 1.
67. Ibid., p. 1.
68. Ibid., p. 14.
69. Ibid., p. 15.
70. Ibid., p. 15.
71. Ibid., p. 22.

72. Ibid., p. 20.
73. Ibid., p. 12.
74. Victor E. Frankl, *Man's Search for Meaning* (New York: Pocket Books, 1963).
75. Siegel, p. 33.
76. Ibid., p. 34.
77. Ibid., p. 3.
78. Ibid., p. 8.
79. Ibid., p. 8.
80. Mihaly Csikszentmihalyi, Ph.D., *Flow: The Psychology of Optimal Experience* (New York: HarperPerennial, 1990).
81. Robert Epstein, Ph.D., *Psychology Today* (Volume 32, No.3), p. 4.
82. A.S. Wellner & D. Adox, *Psychology Today* (Volume 32, No.3), p. 34.

Made in the USA
San Bernardino, CA
27 December 2015